Phlebography of the Lower Limb

Phlebography of the Lower Limb

M. Lea Thomas
MA, MB, BChir, FRCP, FRCR
Senior Physician in Radiology, St Thomas' Hospital and Honorary Lecturer, St Thomas' Hospital Medical School, London.
Clinical Teacher in the Faculty of Medicine, University of London

Foreword by
Michael Hume MD
Professor of Surgery, Tufts University School of Medicine, Boston

CHURCHILL LIVINGSTONE
EDINBURGH LONDON MELBOURNE AND NEW YORK 1982

CHURCHILL LIVINGSTONE
Medical Division of Longman Group Limited

Distributed in the United States of America by
Churchill Livingstone Inc., 19 West 44th Street,
New York, N.Y. 10036, and by associated
companies, branches and representatives
throughout the world.

© Longman Group Limited 1982

All rights reserved. No part of this publication
may be reproduced, stored in a retrieval system,
or transmitted in any form or by any means,
electronic, mechanical, photocopying, recording
or otherwise, without the prior permission of the
publishers (Churchill Livingstone, Robert Stevenson
House, 1–3 Baxter's Place, Leith Walk,
Edinburgh, EH1 3AF).

First published 1982

ISBN 0 443 01841 3

British Library Cataloguing in Publication Data
Lea Thomas, M.
    Phlebography of the lower limb.
    1. Extremities, Lower—Radiography
    2. Pelvis—Radiography 3. Veins—Radiography
    I. Title
    616.1'40757   RC695

Library of Congress Catalog Card Number 81–69845

Typeset by CCC, printed and bound in Great Britain by
Wiliam Clowes (Beccles) Limited, Beccles and London

Foreword

The phlebogram is an incomparable resource for information about the living pathology of veins and is the basis of all new directions in related diagnosis and treatment. In this time of non-invasive diagnosis, the phlebogram is still regarded as the 'gold standard'. A truly comprehensive exposition of phlebography is as welcome as it is timely, therefore.

The phlebogram is not made obsolete by new techniques for demonstrating venous thrombosis, though it can safely be omitted if two (occasionally one) of the less-invasive tests are unambiguously normal. Recent refinements of therapy present many options for treatment and the correct one can only be selected by reference to a phlebogram. One needs to know more than whether thrombosis is present. How recent, how extensive is thrombosis, what is the risk of embolism—answers to such questions and much more information are needed for appropriate treatment. Complications of treatment with anticoagulant drugs are more serious and more frequent than the complications caused by a phlebogram. The writer remains persuaded that precise diagnosis by phlebography should be omitted only under very unusual circumstances.

Thrombolytic therapy can be recommended with increasing confidence for extensive venous thrombosis, but it is difficult to imagine an experienced clinician who would begin this complicated therapy without having a phlebogram. One seeks evidence that thrombosis is of recent duration and so extensive that it is unreasonable to expect an equally good result with anticoagulants alone.

New treatment of venous disease will always increase the indications for phlebography, either for case selection or evaluation of results. Reconstructive surgery of the veins, either bypass of chronic obstruction or to restore competence to major venous valves will require both physiologic testing and the phlebogram. Some phlebographic techniques undoubtedly will change in a parallel enterprise to the improvements now being sought in surgical techniques. The period of slow progress in reconstructive surgery of the veins is giving way to a quickening of interest noticeable today. Many patients now obliged to wear elastic stockings might be managed better by a truly effective operation to improve the competence of major venous valves. Functional phlebography will demonstrate the effectiveness of all stages of development of these new directions in vascular surgery.

Plainly the phlebogram remains central to an understanding of diseases of the veins of the lower limb, to their proper management and for new solutions to old problems. By using the right technique its complications can be largely avoided.

No other methods convey as much information with so little special equipment. Why then does the phlebogram remain, for some, less highly regarded than the arteriogram? The only respectable answer is the one offered in the introduction that follows. Quite simply, the phlebogram is more difficult than arteriography both in the technique required for complete display of the vascular anatomy and for full interpretation of the findings. The practical exposition now at hand of a monumental personal experience must surely elevate the quality of vascular diagnosis among its readership, and give encouragement where interest is lacking because of the burden of inferior technique.

Even as the phlebogram is the acknowledged 'gold standard' of diagnosis against which less invasive tests are to be compared, so this volume appears as a gold standard of reference for all who are concerned with the veins of the lower limb.

Boston 1982   Michael Hume

Preface

Despite the frequency of venous disease of the lower limb, there is, as far as I am aware, no work in the English language devoted exclusively to phlebography. Several excellent monographs have been published in other languages and chapters are devoted to phlebography in standard textbooks of angiography.

In writing this book it is my hope that a gap in the literature will be filled and that it will assist both radiologists and clinicians in their day to day practice.

This monograph is based on a personal experience of over 5800 phlebograms of the lower limbs and pelvis. I have restricted the text to contrast phlebography, although many other techniques have been developed in recent years for the study of veins and their pathology. The introduction of these newer methods has necessarily involved a reappraisal of the place of phlebography in the investigation of venous diseases. One of the criticisms levelled against phlebography is that it is an invasive procedure. It is true that some of these newer techniques are non-invasive but many of them are almost as invasive as contrast phlebography. Phlebography requires only a venepuncture followed by the injection of a relatively safe and well tolerated contrast medium, the most unpleasant part being the venepuncture itself, which is a feature of many other methods of investigating the venous system.

Phlebography has one major advantage over other techniques in that it requires no specialised equipment other than that available in all departments of radiology. Many techniques involve expensive equipment and specialised rooms which are frequently not available in a general hospital.

The abnormalities of the deep venous system chiefly involve the deep veins, and are mainly the result of present or past deep vein thrombosis. The technique of phlebography has however a much wider field of application and this scope, I trust, will be apparent in the monograph.

Phlebography has always seemed to me to be the poor relation of arteriography. This is probably because it is more difficult to carry out and interpret. In arteriography the contrast enters the origin of a branching system and its flow demonstrates the vascular territory of that artery. By contrast, the injection in phlebography is made into one of the tributaries of the venous watershed and to obtain optimum results requires a variety of techniques which include the use of tourniquets and posture to control or interrupt venous flow to achieve uniform venous filling. Furthermore, compared with the arterial system, the venous embryology is extremely complicated resulting in a vast number of channels and innumerable anatomical variants which must not be confused with venous pathology.

In writing this monograph it gives me great pleasure to acknowledge the help of many colleagues. I have had the stimulating experience of working for many years in close cooperation with Professor Norman Browse, Mr F. B. Cockett, Professor J. B. Kinmonth and in later years with Mr David Negus. The experience, help and encouragement of these colleagues, who between them have contributed so much to the understanding of venous problems, has been the main stimulus for my deciding to commit my experiences to paper.

Others from whom I have learnt a great deal, either in personal conversation or through their publications include Dr Robert May of Innsbruck, Dr H. E. Schmitt of Basel and Dr Michael Hume of Boston. Dr Hume has kindly written the foreword and for this compliment I am most appreciative.

Over the 15 or more years in which I have been interested in phlebography I have been assisted by, and received constructive criticism from a large number of my junior staff, too many to name individually. To these too I extend my thanks.

I am also indebted to the many non-medical staff who have assisted me, in various ways, with the examinations.

The reproductions which form such an important part of this book are the work of Mr T. W. Brandon and his staff in the Photographic Department of St Thomas' Hospital who have taken endless trouble preparing illustrations of the highest quality. The line drawings have been carried out most skilfully by the Department of Medical Illustration.

In the preparation of this book I should like to record my thanks particularly to Dr Greg Briggs, Dr Nicholas Perry and Dr Huw Walters who have read the manuscript and offered valuable suggestions and criticisms which have greatly improved my original concept. My secretary Miss Linda Lewis has typed the entire manuscript and to her I am greatly indebted.

Finally I wish to record my grateful thanks to my publishers Churchill Livingstone.

London, 1982   M.L.T.

To my colleagues, N.L.B., F.B.C. and J.B.K.

Contents

1.	Introduction	1
2.	Anatomy and physiology	6
3.	Techniques	25
4.	Complications	54
5.	Artefacts	66
6.	Thromboembolism	85
7.	Post thrombotic states	116
8.	Varicose veins	137
9.	Malformations	153
10.	Extrinsic compression	180
11.	Miscellaneous uses of phlebography	199
	Index	217

1

Introduction

HISTORICAL ASPECTS

The first phlebogram to be undertaken in a living person was performed by Berberich and Hirsch in 1923[6] who injected strontium bromide into an arm vein. McPheeters and Rice[27] produced a phlebogram using lipiodol, but the method with its attendant dangers, was little used. The next major advance in phlebography came with the introduction of the relatively well tolerated water soluble di-iodinated pyridine derivatives such as iodopyracet and iodomethamate.[13, 30, 35]

Frimann-Dahl in 1935[14] described the first contrast demonstration of fresh venous thrombus, occurring in the femoral vein. Three years later Dos Santos[11] published his studies on a large number of patients and obtained the first wide acceptance of phlebography. His method involved the injection of contrast medium into a superficial vein behind the lateral malleolus, then taking a series of films as the contrast ascended the leg – the technique now known as ascending phlebography. This particular method of phlebography however, filled both the superficial and deep veins making film interpretation difficult.

In 1941 Lindblom[25] introduced a supra-malleolar inflated cuff to encourage selective filling of the deep veins. He also minimised the layering effect due to the hyperbaric nature of contrast medium by examining patients in the semi-erect position. The use of a tourniquet to prevent filling of the superficial veins and assist filling of the deep veins was then adopted and popularised by Bauer.[5] In 1959 May and Nissl[26] published their method of phleboscopy, which in contrast to the previously used blind over-couch exposures, enabled them to follow the flow of contrast medium on the fluoroscope taking 'spot' films to record the event.

The Scandinavians have always been pre-eminent in the field of phlebography. Greitz,[15] Gullmo,[18] Almen and Nylander,[2] and Haeger[20] not only improved the technique but produced standard criteria for the interpretation of phlebograms. Bauer[4, 5] made the most outstanding contribution in the uses of phlebography and his papers and monographs, both in their detail of technique and interpretation, remain classics to this day.

Other workers who have made notable contributions to phlebography include: in France, Leriche,[24] Servelle,[36] and Olivier;[29] in Austria, May and Nissl;[26] in Germany, Hach;[19] and in the United States of America, De Weese and Rogoff.[10]

In Great Britain Dow[12] and Gryspeerdt[17] prepared the way for phlebography in this country.

In recent years the work of Schobinger,[33, 34] Schmitt[32] and Albrechtsson and Olsson[1] are especially noteworthy.

INDICATIONS FOR PHLEBOGRAPHY

The various indications for phlebography and the type of investigation which should be carried out are described more fully in Chapter 2 and subsequent chapters. The main indications may be summarised as follows:

To show if the deep venous system is morphologically normal
If a correct technique is used, filling of the three paired stem veins of the calf, the popliteal, the femoral, the external, internal and common iliac veins and the lower inferior vena cava should result in every case.[15, 21, 22] Valves should be shown in the veins of the legs, their demonstration being assisted by the semi-erect posture and the use of the Valsalva manoeuvre (Ch. 2).

To demonstrate venous obstruction
This may be due to the presence of recent venous thrombosis (Ch. 6) or to post thrombotic obstruction where recanalisation has not occurred (Ch. 7). In addition the thin walled veins are easily deformed, compressed and obstructed by neighbouring normal or abnormal structures (Chs. 5, 10).

To demonstrate incompetent communicating veins
It is not unusual for contrast to pass retrogradely through competent communicating veins and it is for this reason that some authors[21, 31] regard phlebography as an unreliable method to demonstrate the presence of incompetent communicating veins. This view is not shared by the author. Provided a correct technique is used, and the additional signs of dilatation, absence of valves and the connection of an incompetent communicating vein with varicose veins is seen, a high degree of accuracy in the interpretation of significant retrograde flow can be obtained.[23] The technique is described in Chapter 2 and the interpretation of the phlebogram in Chapter 8.

To investigate the cause of recurrent varicose veins
Unexpected recurrence of varicose veins following surgery or sclerotherapy usually indicates an unsatisfactory therapeutic measure. The commonest cause is an inadequate stripping of superficial veins or where a small incompetent communicating vein has been missed. Occasionally additional incompetent communicating veins develop after apparently successful treatment. Phlebography is directed to the demonstration of these incompetent communicating veins and to show the state of the deep venous system.

The investigation of venous ulceration
Venous ulceration is not usually associated with primary varicose veins but is part of the post thrombotic syndrome (Ch. 8). Phlebography is performed to

demonstrate any post thrombotic changes, the patency of the deep venous system and the localisation of incompetent communicating veins.

The investigation of the swollen leg
A swollen leg may be due to a number of causes including cellulitis and lymphoedema. It may also be caused by venous incompetence or obstruction, and phlebography is concerned with the exclusion of the latter two possibilities.

The diagnosis and management of deep vein thrombosis.
Phlebography will indicate the size, position and the liability of a thrombus to embolise. It is thus an essential prerequisite not only to confirm the presence of deep venous thrombosis where it has a diagnostic accuracy of about 95 per cent,[8] but in deciding the future management whether by conservative measures such as anti-coagulation, or by surgery. Repeat phlebography is sometimes indicated to determine the success or otherwise of therapy instituted (Ch. 6).

The detection of deep venous thrombosis in establishing the source of pulmonary emboli
The presence of venous thrombosis in the legs together with a strong clinical history or other data suggesting pulmonary embolism provides good presumptive evidence of thromboembolism and the likely source. Pulmonary embolism can be confirmed by pulmonary angiography or ventilation – perfusion isotope scanning.

Complete phlebography of the lower limb to demonstrate the veins from the foot to the lower inferior vena cava will reveal, in the majority of cases, the source of pulmonary embolism. It will also demonstrate residual thrombus which may be a further source of embolic episodes (Ch. 6).

The investigation of venous malformations
Phlebography helps in the diagnosis and management of venous dysplasias. Ascending phlebography will indicate the state of the deep venous system; direct injections into the venous malformation will demonstrate its connections with the deep and superficial veins and arterial injection with follow-through into the venous phase (arteriophlebography) will indicate the size and extent of the lesion (Ch. 9).

Functional studies
Dynamic phlebography, which combines a phlebographic study with exercising of the calf musculature,[2, 3, 10] may be used to examine directly the functioning of the calf muscle pumps. Nowadays other methods such as isotope studies and physiological studies are used. Phlebography together with venous pressure measurements[28, 37] can be usefully combined to elucidate various aspects of venous function (Ch. 7).

Extrinsic obstruction
Veins have thin walls and are easily deformed by neighbouring structures, thus phlebography may play a part in the investigation of mass lesions which may compress the venous system. The walls of the veins, unlike those of arteries, are

also well supplied with lymphatics so that tumour invasion may also be demonstrated by phlebography.

LIMITATIONS OF CONTRAST PHLEBOGRAPHY

A disadvantage of contrast phlebography is that it is not a simple bed-side procedure. Some patients find venepuncture, particularly in the foot, unpleasant and though pain or discomfort from the injection of the conventional high osmolality contrast media occurs it is rarely intolerable.

While most parts of the venous system can be fully demonstrated by phlebography, Cotton and Clark[9] showed with the use of corrosion casts that there is a vast network of muscle veins in the calf which are quite impossible to fill completely by any radiographic technique.

There are other sites where difficulties may be encountered. The profunda femoris vein fills completely in only one third of patients. This occurs when there is a loop connection between this vein and the superficial femoral vein. The internal iliac veins can only be fully filled by intra-osseous phlebography. The left common iliac vein is more difficult to opacify than the right because of compression by the overlying right common iliac artery but this can be overcome by the bolus technique (Ch. 3).

ALTERNATIVE METHODS OF INVESTIGATING THE VENOUS SYSTEM

In recent years a number of non-invasive diagnostic tests have been developed for the investigation of venous system. These include the Doppler flow detector technique, plethysmography, thermography, isotope phlebography and foot volumetry.

These methods are not discussed in this book and interested readers are referred to the original sources and to the publication edited by Bernstein.[7]

Generally these newer methods are useful as screening tests and are complementary to contrast phlebography.

CONCLUSION

Despite the introduction of new and often non-invasive techniques for investigation of the venous system the demand for phlebography continues to increase. This is because these methods are often not as accurate and do not give as much information as radiographic contrast phlebography. It would appear that phlebography will remain the final arbiter on the anatomical and pathological state of the venous system for the foreseeable future.

Contrast phlebography does have some disadvantages and complications which will be discussed later, but is generally simple to perform, requires no specialised equipment, is well tolerated by patients and gives information not obtainable by any other method.

REFERENCES

1. Albrechtsson U, Olsson C G 1976 Thrombotic side effects of lower limb phlebography. Lancet 1: 723
2. Almen T, Nylander G 1962 Serievenographien des normalen Unterschenkels wahrend muskularer Krontraktien und Erschlaffung. Acta Radiol 1: 345
3. Anroldi C C, Greitz T, Linderholm H 1966 Variations in cross-sectional area and pressure in the veins of the normal human leg during rhythmic muscular exercise. Acta chir scand 132: 507
4. Bauer G 1940 A venographic study of thrombo-embolic problems. Acta chir scand 84: Suppl 61
5. Bauer G 1942 A roentgenological and clinical study of the sequels of thrombosis. Acta chir scand 84: Suppl 74
6. Berberich J, Hirsch S 1923 Die rontgenographische darstellung der arterien and venen ani lebenden menschen. Klin Wschr 2: 2226
7. Bernstein E F 1978 (ed) Non invasive diagnostic techniques in vascular disease. The C V Mosby Co, Saint Louis
8. Browse N 1978 Diagnosis of deep vein thrombosis. Brit Med Bull 34: 163
9. Cotton L T, Clark C. 1965 Anatomical localisation of venous thrombosis. Ann Roy Coll Surg Engl 36: 214
10. De Weese J A, Rogoff S M 1958 Clinical uses of functional ascending phlebography of the lower extremity. Angiology 9: 268
11. Dos Santos J C 1938 La phlebographie directe. Conception, technique, premier resultats. J int Chir 3: 625
12. Dow J D 1951 Venography of the leg with particular reference to acute deep thrombophlebitis and to gravitational ulceration. J Fac Radiol 2: 180
13. Edwards E A, Biguria F 1934 A comparison of Skiodan and Diodrast as venographic media, with special reference to their effect on blood pressure. New Engl J Med 211: 589.
14. Frimann-Dahl J 1935 Prospektive Röntgenuntersuchungen Act chir scand Suppl 36
15. Greitz T 1954 The technique of ascending phlebography of the lower extremity. Acta Radiol 42: 421
16. Greitz T 1955 Phlebography of the normal leg. Acta Radiol 44: 1
17. Gryspeerdt G L 1953 Venography of the lower limb. Brit J Radiol 26: 329
18. Gullmo A 1964 Periphere venen. Handbuch der Med Radiol Band X
19. Hach W 1976 Phlebographie der Bein – und Beckenvenen. Schnetztor Verlag, Konstantz
20. Haeger K 1966 Venous and lymphatic diseases of the leg. Bokforlaget Universitat Och Skola Lund, Sweden
21. Højensgard I C 1951 Kronik venos insufficiens in under extremiteterne. Forst Hansen, Kopenhagen.
22. Lea Thomas M 1972 Phlebography. Arch Surg 104: 145
23. Lea Thomas M, McAllister V, Rose D J, Tonge K 1971 A simplified technique of phlebography for the localisation of incompetent perforating veins of the legs. Clin Radiol 23: 86
24. Leriche R, Servelle M 1943 Phlebographie dans les phlebites. Mem Acad Chir 69: 313
25. Lindblom K 1941 Phlebographische Untersuchungen des Unterschenkels bei Kontrastinjektion in eine subcutane vene. Acta Radiol 107: 136
26. May R, Nissl R 1959 Die phlebographie der unteren extremitat. Thieme, Stuttgart
27. McPheeters, H O, Rice C O 1929 Varicose veins – the circulation and direction of the venous flow. Surg Gynec Obstet 49: 29
28. Negus D 1970 The post-thrombotic syndrome. Ann Roy Coll Surg Engl 47: 92
29. Olivier C L 1961 Les Thromboses anciennes de veins iliaque primitives et externs. Presse med 45: 1753
30. Ratschow M 1930 Uroselektan in der Vasographie unter spezieller Berucksichtigung der Varikographie: Forschr Rontgenstr 42: 37
31. Renes G J 1950 Over de oozaken het underzoek en de benhandeling van spataderen van de onderste ledermaten. Thesis, Leiden
32. Schmitt H E 1977 Aszendierend phlebographie bei tiefer venethrombose. Verlag Hans Huber, Bern
33. Schobinger E 1960 Intraosseous venography. Grune and Stratton, New York
34. Schobinger R A 1977 Periphere Angiodysplasien. Verlag Hans Huber, Bern
35. Schwarz E 1934 Die krampfadern der unteren extremitat mit besonderer Berucksichtigung ihrer Entstehung under Benhandlung. Ergebn Chir Orthop 27: 256
36. Servelle M 1944 La Venographie. Maloine, Paris
37. Van Der Hyde M N (1961) Phlebography and venous pressure determination. H E Stenfert Kroese, N V – Leiden

2

Anatomy and physiology

INTRODUCTION

A knowledge of the normal anatomy of the venous system is essential for the interpretation of phlebograms.

The anatomy described in this chapter is based on that shown by phlebography; important applied anatomical features of interest in the pathogenesis and treatment of venous diseases are discussed where appropriate.

THE VEIN

Structure

The veins are in direct continuity with the arteries and have so many similarities in general structure and function that it might be assumed that diseases of the veins would be similar to those of the arteries. There are however, important differences which reflect the different diseases found in the venous system. These are: (1) that veins have valves, (2) they have lymphatics, (3) there is a relative paucity of elastic tissue and muscle in their walls and (4) they are part of a low pressure system. The result is that the most common venous diseases are thrombophlebitis which may result from the lymphatics allowing bacteria to enter the vein wall, phlebothrombosis due to stasis, and varicose veins as a result of distension.[1]

The walls of the veins, like those of the arteries are composed of three coats: the tunica intima, the tunica media and the tunica adventitia. The main difference between arteries and veins lies in the comparative weakness of the muscular layer, especially in the media and the much smaller proportion of elastic tissue.

In the smallest veins these coats are difficult to distinguish. The media varies considerably between veins of different calibre. In the venules it is thin and composed almost exclusively of smooth muscle. In the medium sized veins it consists of a thick layer of connective tissue with elastic fibres and some smooth muscle fibres usually arranged circumferentially. The larger veins, including the upper femoral and iliac veins have a much reduced amount of smooth muscle in the tunica media, and in the inferior vena cava it is almost entirely absent.

The muscle fibres in the main superficial veins of the leg have considerable contractile power, but the media is less well developed in the smaller tributaries of these veins which are more liable to dilate and become tortuous and varicose in response to sustained intravascular pressure.

Valves

Unlike arteries, veins posses valves which direct blood flow towards the heart. The valves have two leaflets consisting of folds of intima reinforced centrally with connective tissue, the free edge of the leaflets lie in the direction of the blood flow.

The common iliac veins and the inferior vena cava are almost always valveless. Valves are occasionally found in the external iliac veins and usually in the main trunk of the internal iliac vein and its tributaries. There are many valves in the distal veins of the limb but these become progressively fewer in number in the more proximal veins such as the superficial femoral and profunda femoris veins (Fig. 2.1).

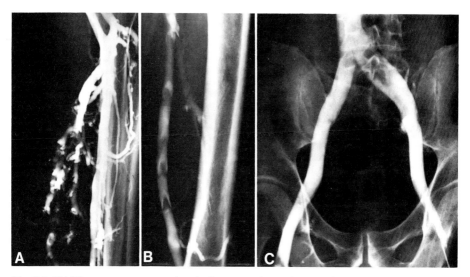

Fig. 2.1 (A) There are numerous valves in the stem and muscle veins of the calf. (B) There are fewer valves in the more proximal veins. In this superficial femoral vein the bicuspid valves are clearly shown because a Valsalva manoeuvre is being performed. (C) The external and common iliac veins are usually valveless. Valves are present in the internal iliac veins, preventing retrograde filling.

There are no valves in the baggy sinusoidal veins of the soleal muscles (Fig. 2.2) but the venous arcades which also drain the soleal and gastrocnemius muscles have numerous valves.[11]

All the communicating veins in the lower part of the calf, and all the veins connecting the deep and superficial venous systems of the lower limb have valves which ensure that blood can only pass from the superficial to the deep system (Fig. 2.3).

There is some doubt about the significance of the valves in the communicating veins of the feet; it seems likely that they are only partly valved so that blood can flow from the feet into both the deep and the superficial venous systems.[6] This is probably the reason why an ankle tourniquet is helpful to fully demonstrate the deep veins of the calf. As a consequence of this bidirectional flow in the foot, in

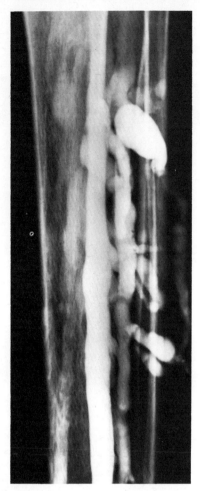 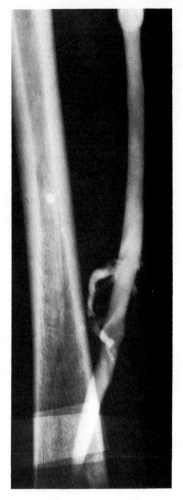

Fig. 2.2 Typical baggy, valveless soleal muscle veins.

Fig. 2.3 A competent valve is present in an adductor canal communicating vein.

contrast to the situation in the rest of the leg, veins of the foot have no real haemodynamic importance.

VEINS OF THE FOOT

The venous drainage of the foot consists of four systems[5]:

1. The superficial dorsal venous arch i.e. the long and short saphenous veins joined together by the arch and its tributaries
2. The plantar cutaneous arch joining the medial and the lateral marginal veins
3. The deep venous system of the sole, i.e. the lateral and medial plantar veins which become the posterior tibial veins
4. The communicating veins which connect the deep and superficial networks (Fig. 2.4).

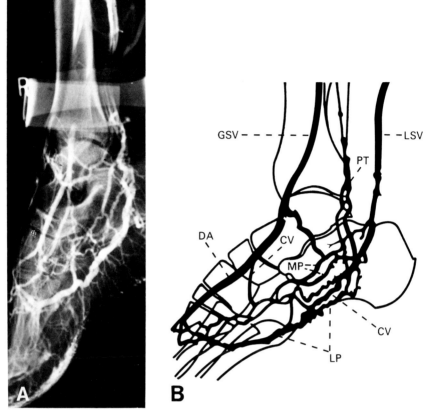

Fig. 2.4 (A) A normal foot phlebogram. (B) Schematic drawing of the foot veins. (GSV – Greater (long) saphenous vein; LSV – Lesser (short) saphenous vein; PT – Posterior tibial veins; DA – Dorsal venous arch (medial limb); MP – Medial plantar vein; LP – Lateral plantar vein; CV – Communicating veins connecting the medial plantar veins with the medial limb of the dorsal venous arch and the lateral plantar veins with the lateral marginal vein.) (After Pegum & Fegan, 1967)[14].

VEINS OF THE LOWER LEG

From an anatomical and physiological point of view, the veins of the lower leg can be divided into four groups (A) the deep leg veins, (B) the muscle veins of the calf, (C) the superficial veins and (D) the communicating veins.

The deep leg veins

The deep veins of the lower leg consist of the three paired stem veins which are venae comitantes accompanying the arteries: the anterior tibial veins, the posterior tibial veins and the peroneal veins. Each vein may divide into several trunks which surround the artery and anastamose freely with each other. The anterior tibial veins drain the blood from the dorsum of the foot and run in the deep part of the extensor compartment close to the interosseous membrane. The posterior tibial veins are formed by the confluence of the superficial and deep plantar veins behind the ankle joint beneath the flexor retinaculum. The peroneal veins lie directly behind and medial to the fibula.

10 PHLEBOGRAPHY OF THE LOWER LIMB

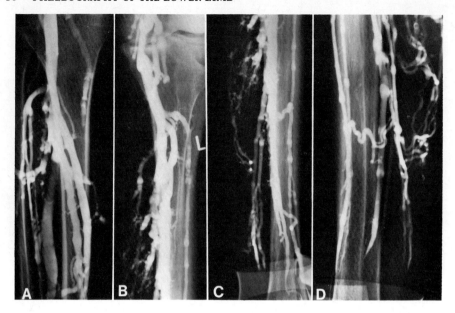

Fig. 2.5 The three paired stem veins of the calf; the anterior tibial, peroneal and the posterior tibial veins. There are frequently communications between the veins. In the upper part of the calf the veins merge into single trunks and then unite to form the popliteal vein. (A) Anterior projection with foot internally rotated. (B, C, and D) Lateral projections with leg externally rotated so that the calf lies flat on the X-ray table. This view gives a better demonstration of the veins (stem and muscle) of the calf. The vein which lies behind the tibia and passes upwards towards it is always the posterior tibial vein.

On an anterior view of a leg phlebogram, with the foot internally rotated to separate the images of the tibia and fibula, the anterior tibial veins lie more laterally, often over the fibula, and the posterior tibial veins lie medially crossing the shaft of the tibia in its lower third as they pass upwards behind the medial malleolus.

The individual veins, particularly the anterior tibial veins, are easier to identify on a lateral view of the phlebogram (Fig. 2.5). The size of the veins varies but the anterior tibial veins are always smaller than the others.

Fig. 2.6 (A) The popliteal and superficial femoral veins. The veins pass medially towards the groin. In phlebograms the popliteal vein becomes the superficial femoral vein where it crosses the medial border of the femur. The common femoral vein is formed by the confluence of the superficial and deep femoral veins below the inguinal ligament. (B) A vena comitans often accompanies the vein. Venae comitantes can function as collaterals in venous obstruction.

Fig. 2.7 In about a third of legs there is a direct connection between the superficial and deep femoral veins so that complete filling occurs from ascending phlebography.

ANATOMY AND PHYSIOLOGY 11

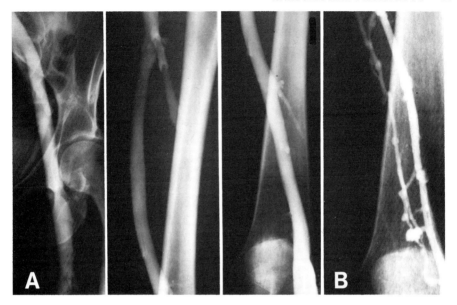

Fig. 2.6

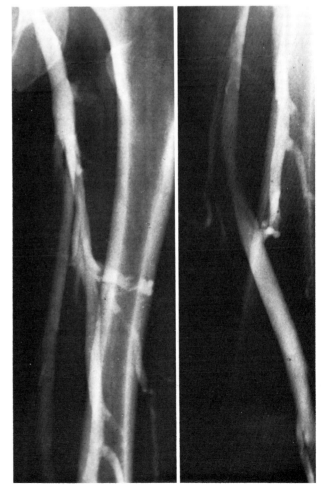

Fig. 2.7

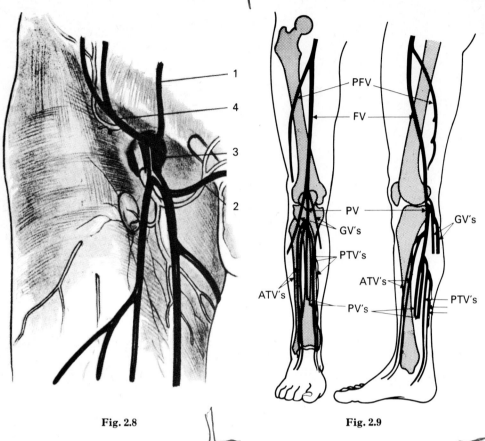

Fig. 2.8 Tributaries of the common femoral and/or external iliac vein. (1) Superficial epigastric vein (2) Superficial external pudendal vein (3) Deep external pudendal vein (4) Superficial circumflex iliac vein (After May & Nissl, 1959)[9].

Fig. 2.9 The main deep veins of the leg. (PV's – Peroneal veins; ATV's – Anterior tibial veins; PTV's – Posterior tibial veins; GV's – Gastrocnemius veins; PV – Popliteal vein; FV – Superficial femoral vein; PFV – Profunda femoris vein) (After May & Nissl, 1959)[9].

In the upper part of the calf the paired stem veins merge into single trunks and then unite at differing levels to form the popliteal vein. In about 50 per cent of patients the confluence is distal to the knee joint with a similar number uniting above the knee joint.

The superficial femoral vein is the direct continuation of the popliteal vein, and passes obliquely upwards and medially across the lower third of the femur through the adductor canal towards the groin (Fig. 2.6). Below the inguinal ligament it receives the profunda femoris (deep femoral) vein which is often not fully demonstrated by phlebography. In the author's experience a third of phlebograms show a direct connection between the superficial femoral and the profunda femoris veins in the lower third of the thigh. In these circumstances the whole of the profunda femoris vein fills during ascending phlebography (Fig. 2.7).

ANATOMY AND PHYSIOLOGY 13

The common femoral vein is formed by the confluence of the superficial femoral and profunda femoris veins. This vein has a number of tributaries which are important in femoral, iliac and inferior vena caval obstruction. These tributaries include the superficial circumflex iliac vein, the superficial epigastric vein, the superficial and deep external pudendal veins and the long saphenous vein (Fig. 2.8).

The main stem veins of the leg are shown diagrammatically in Figure 2.9.

The muscle veins of the calf

The veins draining the soleal muscle form arcades joining the posterior tibial and peroneal veins (Fig. 2.10). The soleal sinusoidal veins (see Fig. 2.2) are dilated segments of these venous arcades.

The gastrocnemius muscle on the other hand is drained by two or more thin walled, valved veins which join the popliteal vein at varying levels in the popliteal fossa (Fig. 2.11).

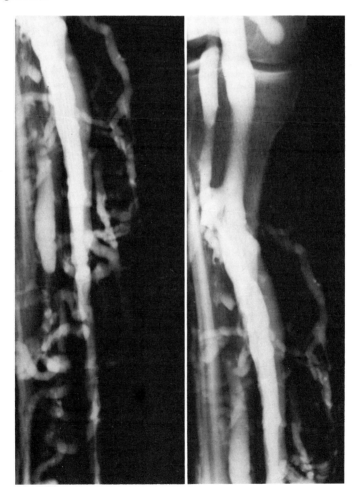

Fig. 2.10 Venous arcades joining the posterior tibial veins and the peroneal veins.

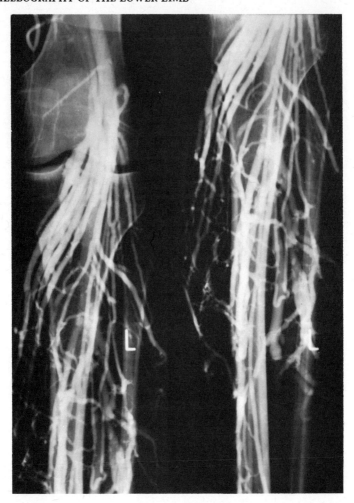

Fig. 2.11 Gastrocnemius veins. These tend to be multiple, straight and are valved. They join the upper popliteal vein. The soleal veins are baggy and valveless (See Fig. 2.2).

The superficial veins

The long saphenous vein is formed by the union of the veins from the medial part of the sole of the foot and the medial marginal vein. It runs upwards in front of the medial malleolus, across the length of the antero-medial aspect of the limb to join the common femoral vein at the groin.

At the ankle the position of the long saphenous vein is fairly constant lying in the groove at the anterior border of the medial malleolus. Thus, if surgical exposure of a vein for phlebography is required this vein can easily be identified in the presence of considerable oedema.

The long saphenous vein can easily be demonstrated on phlebography by direct injection into it (Fig. 2.12). The vein receives a number of tributaries throughout its course, the most important of which are the posterior arch veins of which the medial communicating veins are tributaries (Fig. 2.13).

ANATOMY AND PHYSIOLOGY 15

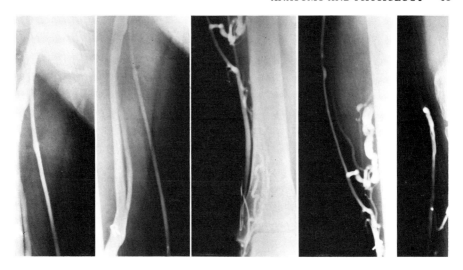

Fig. 2.12 The course of the long saphenous vein shown by injection of the foot vein with the patient horizontal and without an ankle tourniquet.

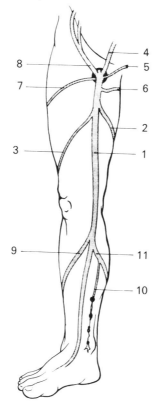

Fig. 2.13 The superficial veins of the legs and their connections. (1) Long saphenous vein (2) Medial accessory saphenous vein (3) Lateral accessory saphenous vein (4) Superficial epigastric vein (5) External pudendal vein (6) Medial circumflex femoral vein (7) Lateral circumflex femoral vein (8) Superficial circumflex iliac vein (9) Anterior vein of the leg (10) Posterior arch vein (11) Anastamosis with the short saphenous vein (After May, 1979)[8].

16 PHLEBOGRAPHY OF THE LOWER LIMB

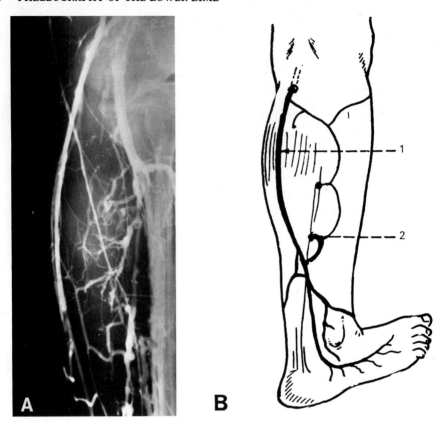

Fig. 2.14 The short saphenous vein. (A) Ascending phlebogram. (B) Diagram. (1) The short saphenous vein (2) The large constant ankle communicating vein (After Dodd & Cockett, 1976)[2].

The short saphenous vein begins at the outer border of the foot behind the lateral malleolus and is formed by the union of the lateral marginal veins with numerous small veins draining the outer aspect of the heel. It first passes upwards on the lateral border of the tendo Achillis and then along the middle of the back of the calf (Fig. 2.14). In the lower third of the calf it lies on the deep fascia; in the middle third it lies in the fascial layer covering the gastrocnemius muscle. Its exact termination is extremely variable. Most commonly it unites with the popliteal vein in the popliteal fossa a few centimetres above the knee joint, but it may join the upper part of the popliteal vein, or the deep veins of the calf, or continue upwards to join the femoral vein. Its exact site of termination can be established by superficial phlebography of the leg in the lateral position. This is sometimes important if surgical ligation is being contemplated in the treatment of varicose veins.

The short saphenous vein chiefly drains the lateral aspect of the heel and calf. At the lower end of the calf it is joined by an inconstant ankle perforating vein which winds around the fibula and connects it to the peroneal vein; it also communicates with the sinusoidal veins of the calf and the long saphenous vein.

The communicating veins

The largest communicating veins are the terminations of the long and short saphenous veins where they join the deep venous system. These however form only a part of a series of communicating veins.

In the lower calf there are medial and lateral communicating veins which are of considerable surgical importance. The medial communicating veins penetrate the deep fascia and empty directly into the lower tributaries of the posterior tibial vein. There are three constant medial communicating veins, one just behind the tip of the medial malleolus and the others approximately three fingers width apart behind the posteromedial border of the lower third of the tibia. These veins are linked to each other forming an arcade known as the posterior arch veins and do not drain directly into the long saphenous vein (Fig. 2.15).

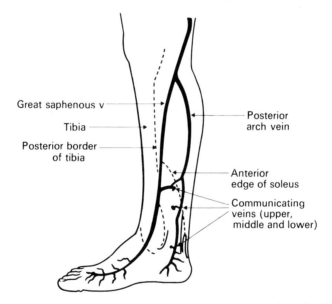

Fig. 2.15 The sites and superficial connections of the medial communicating veins. Note that the three main communicating veins are not directly connected to the long (great) saphenous vein and are not affected by stripping operations.

A lateral communicating vein with a constant position is found at the junction of the lower and middle third of the calf; it joins the short saphenous and the peroneal veins. A further posterior mid calf communicating vein links the short saphenous vein and the soleal sinusoidal veins.

In the thigh there is a constant long communicating vein which joins the long saphenous vein, or one of its tributaries, to the superficial femoral vein in the lower third of the sub-sartorial canal. Incompetence of the vein can be shown by ascending phlebography combined with a Valsalva manoeuvre.

While these communicating veins are relatively constant in position there are many others which have an inconstant position and these too may become incompetent (Fig. 2.16). It is for this reason that accurate localisation by phlebography is often required.

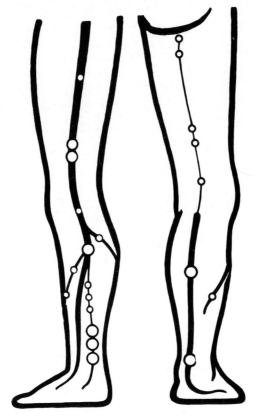

Fig. 2.16 The common sites of the communicating veins. The sites of those marked on this figure are relatively constant, but there are many communicating veins at other sites. (After May, 1979)[8].

THE ILIAC VEINS AND LOWER INFERIOR VENA CAVA

To the vascular surgeon and the radiologist, the external iliac veins, the common iliac veins and the lower inferior vena cava are most important. The internal iliac veins which drain the pelvic viscera are less important, but they may be the site of venous thrombosis and they may also provide a collateral circulation in the presence of venous obstruction (Fig. 2.17).

The external iliac veins
The external iliac veins are the continuation of the common femoral veins. They run from the level of the inguinal ligament to the level of the sacro iliac joints where they are joined infero-medially by the internal iliac veins emerging from the true pelvis (Fig. 2.18).

Phlebographically, since the inguinal ligament cannot be visualised but its level inferred from its known anatomical position, the distinction between the common femoral veins and the external iliac veins is somewhat artificial, and tributaries ascribed to one may in fact drain into the other.

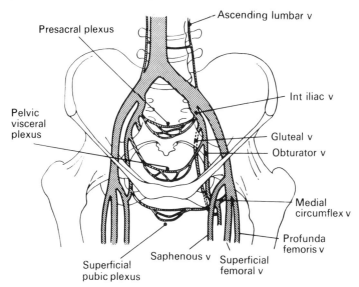

Fig. 2.17 Diagram of potential collateral pathways in parietal pelvic vein obstruction. (After Mavor & Galloway, 1967)[10].

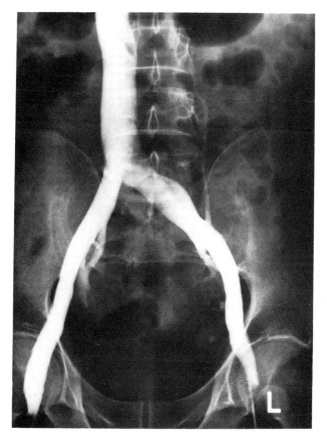

Fig. 2.18 A normal pelvic phlebogram showing the external iliac veins, the common iliac veins and the lower inferior vena cava. The internal iliac veins are seen with a Valsalva manoeuvre as far as competent valves permit.

There is a slight translucency at the termination of the left common iliac vein due to compression by the right common iliac artery in the supine position. It is a normal appearance found in about 50 per cent of phlebograms.

20 PHLEBOGRAPHY OF THE LOWER LIMB

The main tributaries of the external iliac veins are the inferior epigastric vein, the deep circumflex iliac vein and the pubic veins which anastomose with their fellows of the opposite side to form important collateral channels when there is common or external iliac vein obstruction (see Fig. 2.17).

The internal iliac veins

These are formed on the floor of the true pelvis by the union of the gluteal, internal pudendal and obturator veins, and the veins of the pre-sacral and pelvic visceral plexuses. They may form valuable anastomotic channels across the pelvis in common iliac vein obstruction.

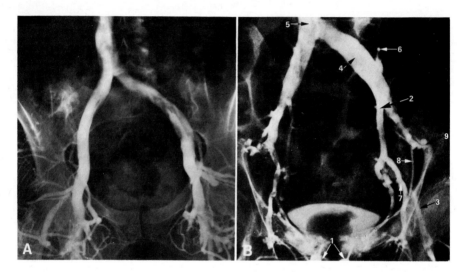

Fig. 2.19 (A) A pelvic phlebogram following bilateral pertrochanteric intraosseous injections in the supine position. The main internal iliac veins are obscured by the external iliac veins and only the gluteal tributaries are seen. (B) Bilateral intraosseous injections into the pubic bones in the supine position. This technique allows visualisation of all the tributaries but is rarely required. (1) Bone marrow needles (2) Internal iliac vein (3) External iliac vein (4) Common iliac vein (5) Inferior vena cava (6) Ascending lumbar vein (7) Obturator vein (8) Internal pudendal vein (9) Gluteal vein.

The veins are best demonstrated phlebographically by the intra-osseous technique with the patient in the supine position allowing the hyperbaric contrast medium to gravitate into these posteriorly situated veins (Fig. 2.19). Alternatively the internal iliac vein may be selectively catheterised as described by Dow.[4]

The internal iliac veins can almost always be demonstrated to some extent by ascending phlebography but the extent will depend on the competence of the venous valves. The technique depends on the Valsalva manoeuvre undertaken during phlebography and when the common iliac veins are filled with contrast medium (see Fig. 2.18).

The common iliac veins

These are short, wide trunks which pass upwards from the level of the sacro-iliac joints to unite on the right side of the fifth lumbar vertebra. The left common iliac

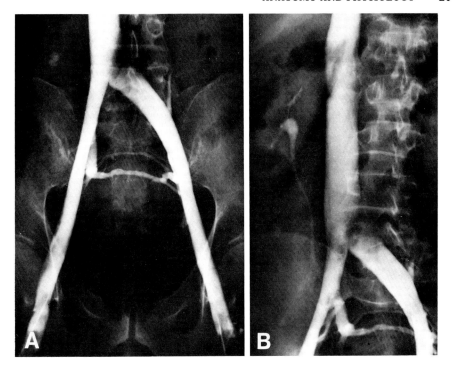

Fig. 2.20 (A) A normal pelvic phlebogram. The lower end of the left ascending lumbar vein can be seen just filling at the junction with the left common iliac vein. Note normal compression defect at the termination of the left common iliac vein. (B) A normal inferior vena cavogram. Non-opacified blood is entering from the right renal vein.

vein is crossed by the right common iliac artery and this causes a variable degree of antero-posterior compression of the termination of this trunk (see Fig. 2.18) which appears as a radiographic filling defect in about 50 per cent of phlebograms. Excessive compression at this site may predispose to thrombosis.[3,13]

The main tributary of the common iliac vein is the ascending lumbar vein which is larger on the left side than the right. It forms a valuable collateral in common iliac and inferior vena caval obstruction through its numerous communications with the vertebral plexuses and the lumbo-azygos venous system.

The lower inferior vena cava

The inferior vena cava ascends from the level of the fifth lumbar vertebra to the right atrium just to the right of the vertebral bodies (Fig. 2.20). The lower inferior vena cava receives a variable number of short, wide lumbar veins which connect with the vertebral plexuses.

PHYSIOLOGY OF THE VEINS OF THE LEGS

Only those aspects of the physiology important to the radiologist are considered here, a more detailed account is available in the monograph by Ludbrook.[7]

The veins are not merely passive conduits with valves so arranged that blood can flow only towards the heart. It is now known that veins can actively contract, propelling blood from the peripheral to the central circulation. The veins form the reservoir of the circulatory system containing up to 70 per cent of the blood in the body, mainly in the lower limbs. In addition to the conduit and capacity functions of the veins of the lower limb, the concept of the so called muscle pump is of great importance.

The deep veins of the calf with the muscles that surround them are enclosed in a rigid fascial sheath. The main stem veins of the calf receive their blood from two main sources; the muscles, whose venous drainage is largely through the sinusoidal veins in the soleal muscle, and from the skin and superficial tissues through the communicating veins which pierce the deep fascia. Valves direct the blood from the superficial veins to the deep veins and from the distal veins to the proximal veins. This arrangement is responsible for the function of the calf muscle pump (Fig. 2.21). In the supine position blood flows evenly along both superficial and deep veins towards the heart.

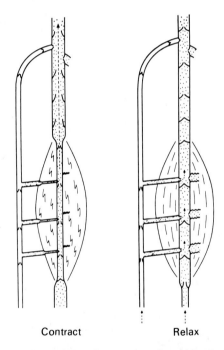

Contract   Relax

Fig. 2.21 Schematic diagram of the calf muscle pump in action. (After Ludbrook, 1966)[7].

In the motionless erect posture the pressure in the dorsal veins of the foot is approximately equal to that of an uninterrupted column of blood extending from the foot to the right atrium. As the calf muscles contract exerting a pressure of over 200 mm Hg a pressure of about 140 mm Hg is measured in the posterior tibial vein and venous blood is pumped from the deep veins of the calf towards the heart. As the muscles relax the venous pressure falls to about 40 mm Hg reflux being then prevented by the closure of the numerous venous valves, and more blood enters the pumping chambers from the muscles and from the superficial

ANATOMY AND PHYSIOLOGY 23

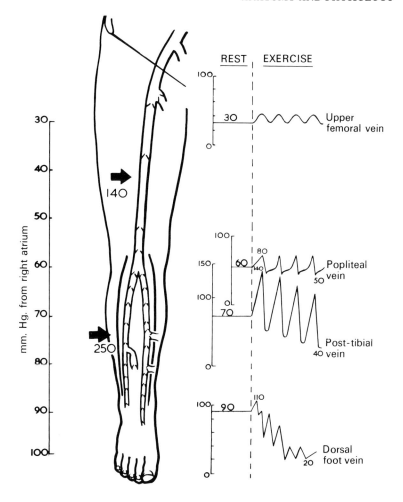

Fig. 2.22 Resting pressure in the deep veins of the legs at various levels in the erect posture and the changes which are produced by the repeated contraction of the calf muscle pump. (After Negus, 1970)[11].

veins through the communicating veins. The result is an increase in the blood flow towards the heart and a fall in the pressure in the distal superficial veins (Fig. 2.22).

In the popliteal and femoral veins, calf muscle exercise produces only minor fluctuations in pressure with little or no change in the mean pressure. The thigh muscle pump contributes much less to venous return.[7,13]

The calf with its dense investing fascia, its unique system of muscle sinusoids, its profuse system of valves and communicating veins appears to be the major pump responsible for venous return from the leg.[13]

Failure of the calf muscle pump may be a sequel of past venous thrombosis and may lead to post thrombotic symptoms which are discussed in Chapter 7.

REFERENCES

1. Boyd W 1970 A textbook of pathology, 8th edn. Lea and Febiger, Philadelphia
2. Dodd H, Cockett F B 1976 The pathology and surgery of the veins of the lower limb. Churchill Livingstone, Edinburgh
3. Cockett F B, Lea Thomas M 1965 Iliac compression syndrome. Brit J Surg 52: 816
4. Dow J D 1973 Retrograde phlebography in major pulmonary embolism. Lancet 2: 407
5. Jacobsen B 1970 The venous drainage of the foot. Surg Gynecol Obstet 131: 22
6. Hach W 1970 Phlebographie der bein – und Beckenvenen. Schnetztor – Verlag, Konstance
7. Ludbrook J 1966 Aspects of venous function in the lower limbs. Charles C Thomas, Springfield, Illinois
8. May R 1979 Surgery of the leg and pelvic veins. Georg Theime Verlag, Stuttgart
9. May R, Nissl R 1959 Die phlebographie der unteren extremitat. Georg Thieme Verlag, Stuttgart
10. Mavor G E, Galloway J M D 1967 Collaterals of the deep venous circulation of the lower limb. Surg Gynec and Obstet 125: 561
11. Negus D 1970 The post thrombotic syndrome. Am Roy Coll Surg 47: 92
12. Negus D 1976 In: Dodd H, Cocket F B (ed) The pathology and surgery of the veins of the lower limb. 2nd Edn Churchill Livingstone, Edinburgh
13. Negus D, Fletcher E W L, Cockett F B, Lea Thomas M 1968 Compression and band formation at the mouth of the left common iliac vein. Brit J Surg 55: 369
14. Pegum J M, Fegan W G 1967 Anatomy of the venous return from the foot. Cardiovasc Res 1: 241

3

Techniques

THE STANDARD TECHNIQUE FOR ASCENDING PHLEBOGRAPHY

The standard technique used by the author is suitable for 95 per cent of clinical situations where phlebography is indicated. The aim of the examination is to demonstrate the deep venous system from the foot to the lower inferior vena cava.

The position of the patient

Ideally, ascending phlebography of the lower limb should be carried out with the patient erect, as in this position the deep venous system always fills provided it is patent and there is always maximum mixing of contrast with blood. However, such a position often requires specialised equipment and the posture is not well tolerated by many patients.

A 20° to 40° tilt is usually all that is required (Fig. 3.1) provided tourniquets are used – steeper table tilts up to 60° may be required in certain clinical situations.

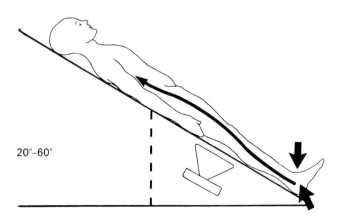

Ascending phlebography

Fig. 3.1 The position used for ascending phlebography (Re-drawn after Ludbrook, 1972)[27].

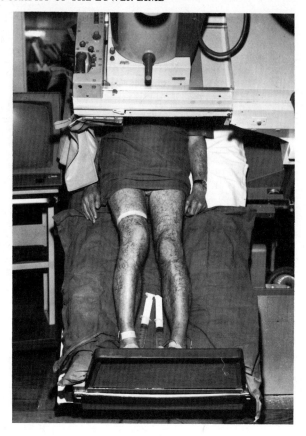

Fig. 3.2 This photograph shows two tourniquets in place and a steep (60°) foot down table tilt. Both legs are internally rotated to separate images of the tibia and fibula.

When the supine position, or only small tilts are used, the layering phenomenon between blood and contrast has to be taken into account but this can be overcome provided sufficient contrast is used, tourniquets applied, and the bolus technique (described later) employed to displace the blood by contrast medium.

The patient should be made as comfortable as possible using hand grips to steady himself where necessary. With steep table tilts the patient is requested to support his weight on the leg not being examined as the muscular contraction due to weight bearing prevents the muscle veins filling.

The leg is internally rotated to separate the images of the tibia and fibula. This not only avoids overlapping of bone, but also separates the veins making them easier to identify (Fig. 3.2).

The vein

Although any vein on the dorsum of the distal part of the foot is suitable for venepuncture, the most constant vein is the medial digital vein of the great toe. A further advantage of this site is that any extravasation can be more easily detected because of the thin layer of skin and soft tissue at the base of the hallux.

Oedema may make the finding of a suitable vein difficult. In these cases firm pressure with the thumb will often disperse oedema from the forefoot sufficiently to make a subcutaneous vein visible. If there is no urgency for phlebography, elevation of the leg for 24 hours will also reduce the oedema. Immersing the foot in warm water, covering the foot with warm packs or sitting the patient on the side of the table with the legs dependent and supported on a chair, are further ways of encouraging venous dilatation before venepuncture. Very rarely, a cut-down onto the vein is necessary to insert a cannula. In this situation it is best to expose the long saphenous vein where it has a constant position in front of the medial malleolus (Ch. 2).

The venepuncture
As the most serious complications of ascending phlebography result from extravasation of contrast medium at the site of venepuncture, a single, clean puncture is desirable. When further venepuncture attempts are necessary, which may be the case in inexperienced hands, contrast may leak into the tissues from earlier puncture sites. The degree of leakage is minimised by leaving needles from failed venepunctures in situ until the examination is finished and where necessary applying pressure until haemostasis is achieved. Contrast is then injected through a successful venepuncture.

The use of a plastic cannula to reduce the frequency of extravasation has been advocated.[14] Venepuncture is undoubtedly more difficult with a plastic cannula

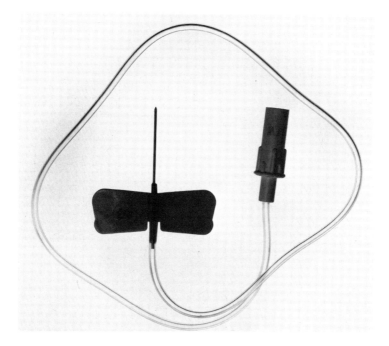

Fig. 3.3 The 'butterfly' needle with connecting tube. The advantages of this needle is that the 'wings' can be gripped for easier introduction and when flat can be taped to the patient's foot to prevent dislodgement.

than with a sharp fine gauge needle, and the advantages of reducing minor extravasation may be outweighed by the distress of multiple venepunctures. Another advantage claimed for the plastic cannula is that it is less likely to be dislodged during the positioning of the patient for lateral views, and during exercise tests to assess the calf muscle pump. If a 'butterfly' needle is strapped in position on the flat part of the dorsum of the foot, dislodgement does not appear to be a problem. The author uses a 21 gauge (or smaller) needle (Fig. 3.3).

The direction of the needle

If the contrast is injected in an upstream direction that is through a needle directed towards the toes with a tourniquet above the ankle (Fig. 3.4), it will tend to pass more directly into the deep venous system through the communicating veins of the foot.[18] The more usually employed downstream injection tends to fill the superficial venous system, the contrast entering the dorsal venous arch which empties into the long and short saphenous systems.

Tourniquets

When contrast medium is injected into a superficial vein of a foot drained by a normal venous system, it will tend to pass into the deep veins of the leg following the normal route of blood flow. However, when the deep venous system is abnormal, satisfactory deep venous filling may be difficult to achieve. Undoubtedly steep table tilts assist the process but a tourniquet above the ankle helps direct the contrast medium into the deep venous system with consequent enhancement of filling and definition of the diseased venous system.[9, 21, 29]

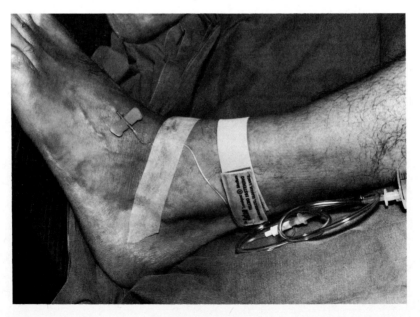

Fig. 3.4 A fine needle is inserted towards the toes after applying a self fastening tourniquet above the ankle.

TECHNIQUES 29

The tourniquet is placed just above the ankle (see Fig. 3.4) and its pressure adjusted during injection to ensure that contrast medium passes up the leg in the deep veins and not exclusively or excessively in the superficial system. When the examination is being carried out for the identification of incompetent communicating veins, the tourniquet has to be sufficiently tight to totally occlude the superficial venous system so that the direction of the flow in the communicating veins can be shown. If an incompetent communicating vein is suspected at the level of the ankle, or when venous ulceration around the ankle prevents the application of the tourniquet at this site, it may be applied around the forefoot.

A tourniquet above the knee (see Fig. 3.2) whilst not essential is valuable in delaying the emptying of contrast from the calf veins and thus improving deep venous filling.

The Valsalva manoeuvre
The Valsalva manoeuvre is performed by asking the patient to take a deep breath,

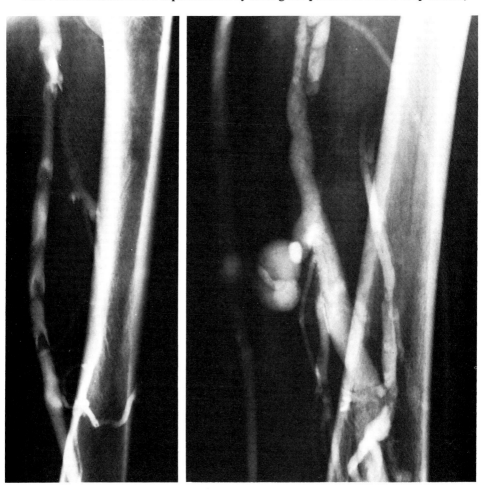

Fig. 3.5 Normal valves shown clearly by the Valsalva manoeuvre.
Fig. 3.6 A Valsalva manoeuvre demonstrates an incompetent communicating vein in Hunter's canal. The deep veins show gross recanalisation changes.

30 PHLEBOGRAPHY OF THE LOWER LIMB

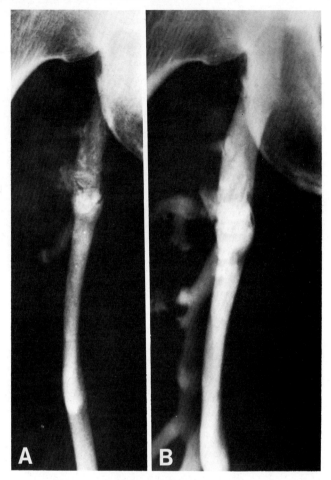

Fig. 3.7 (A) The profunda femoris vein is poorly demonstrated although the superficial femoral vein is well filled. (B) A Valsalva manoeuvre demonstrates the profunda vein much more completely.

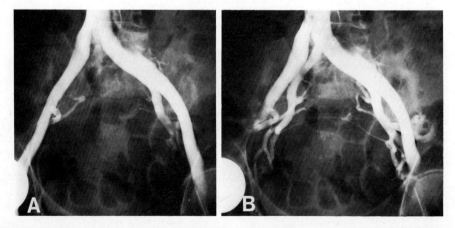

Fig. 3.8 Demonstration of the internal iliac veins. (A) Without a Valsalva manoeuvre. (B) Much improved filling using the Valsalva technique. There is an anatomical variant: both internal iliac veins join on the left.

closing the lips firmly, pinching the nose and then attempting to blow out hard. This forced respiratory effect against a closed glottis causes a rise in intra-thoracic pressure which is transmitted to the inferior vena cava producing reflux of blood towards the limb veins closing the valves if competent. The valve pockets become distended clearly outlining the cusps.

The manoeuvre is useful to confirm the presence of valves (Fig. 3.5), to demonstrate incompetent communicating veins, particularly in Hunter's canal (Fig. 3.6) and to distend the veins to show tributaries such as the profunda femoris (Fig. 3.7) and the internal iliac veins (Fig. 3.8).

The contrast medium
The commonly used contrast media in phlebography are sodium and meglumine diatrizoate, iothalamate or metrizoate.

The author uses meglumine iothalamate 60 per cent (Conray '280') for the legs and the dense sodium iothalamate 70 per cent (Conray '420') for the pelvic veins and the inferior vena cava.

The maximum safe dose of contrast medium is not known but the amount used varies depending on the clinical problem, the necessity for repeat films and the venous capacity of the legs. An average of 50 to 60 ml of contrast medium per leg is usually sufficient but if there is need for supplementary pelvic vein demonstrations, a total of between 200 to 250 ml may be necessary.

No complications from the use of these volumes of contrast medium have been experienced. All the commonly used media give good opacification of the veins but, being hyperbaric, they tend to layer on the dependent side of the vessel and it is therefore essential to completely fill the veins with contrast. The poor quality films of early phlebographers resulted from an inability to use sufficiently large volumes of contrast medium because of toxic side effects. The modern media are remarkably free from serious reactions and larger volumes can now be employed (Ch. 4).

Injection of the contrast medium
Injection of the contrast medium is made by hand using a 50 ml syringe connected to the 'butterfly' needle by a connecting tube about 30 cm long. This is necessary to protect the operator's hands from radiation. Additional protection is provided by a slatted lead rubber shield, or the use of an automatic injector which gives a flow rate of 0.5 ml per second.

With a hand injection the rate of injection is relatively slow, being limited by the small calibre of the needle and the viscosity of the contrast medium. The flow of contrast is continuously monitored by fluoroscopy.

The radiographic equipment
A tilting table is essential. Image intensification with television monitoring and an automatic exposure device are desirable. The use of image intensification combined with television monitoring has considerably improved the diagnostic accuracy of phlebography (Fig. 3.9). It enables the phlebographer to see the direction of flow of the contrast medium, an important aspect when identifying incompetent communicating veins, and to see that the veins are well filled before exposures are made, thus avoiding artefacts (see Chapter 5).

32   PHLEBOGRAPHY OF THE LOWER LIMB

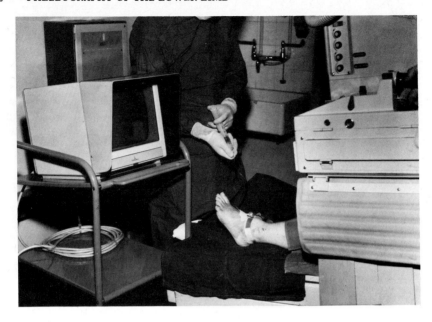

Fig. 3.9 Modern equipment for ascending phlebography. The flow of the contrast medium into the calf is watched on the television monitor. An assistant is using a gentle hand injection. The tourniquet on this leg has been placed below the malleoli to avoid a low medial ulcer (not visible in this photograph). The table is tilted 20° downwards. Slight tilt like this is usually sufficient provided tourniquets are used. The leg is relaxed and not weight bearing, so that contraction of the calf muscles does not occur.

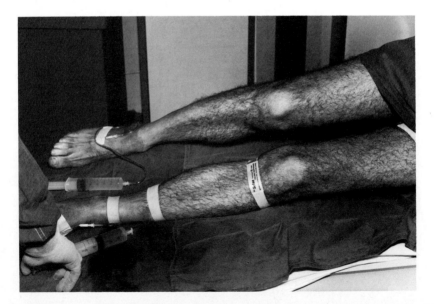

Fig. 3.10 The lateral view is obtained by rotating the leg outwards until the calf lies flat on the table top. An additional tourniquet is being used here to prevent flooding of the superficial veins from an incompetent communicating vein below it.

Small focal spot fluoroscopic X-ray tubes now available produce almost as good radiographic detail as was previously obtained with the long focus-film distance possible with an overcouch tube, and there is now no justification for the use of 'blind' non-fluoroscopic techniques. The quality of the radiographs can be further improved by the use of fluoroscopy equipment with a fixed focus film distance of 1 metre such as is used for remote control barium work. This apparatus is not however generally available.

Views of the leg veins are best recorded on a 35 cm^2 film divided into three.

THE STANDARD TECHNIQUE

Each leg is examined separately. The patient lies supine on the fluoroscopic table and a self fastening tourniquet is applied just above the ankle to distend the foot veins. The 'butterfly' needle is introduced percutaneously into a vein on the dorsum of the foot as described above.

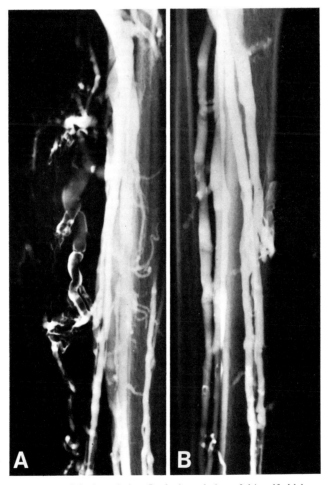

Fig. 3.11 The importance of the lateral view. In the lateral view of this calf phlebogram extensive recent thrombus is shown in the calf muscle veins. (B) In the straight projection it is obscured by the stem veins and the bones.

A second tourniquet is applied above the knee. The table is tilted foot downwards with the leg internally rotated. The patient is instructed not to weight bear on the leg being examined. The needle is connected by an extension tube to a syringe containing the contrast medium and a hand injection is made under television control. The pressure of the ankle tourniquet and the degree of tilting may need to be adjusted to ensure deep venous filling. Films are taken of the deep venous system in the postero-anterior and lateral views. The lateral view is obtained by rotating the leg outwards until the side of the calf lies flat on the table top (Fig. 3.10). In this way a single undercouch tube can be used for both projections (Fig. 3.11). Rotation of the limb may be used to distinguish between the superficial and deep veins (Fig. 3.12). The ankle and thigh tourniquets are then removed in turn and further exposures made to show the veins as far as the inferior vena cava. Firm pressure is applied to the calf muscles in order to empty the veins of contrast so that a bolus enters the more proximal veins just before the exposure is made (Figs. 3.13 and 3.14).

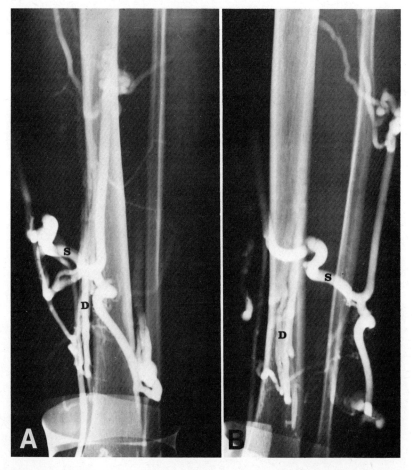

Fig. 3.12 (A) In this straight projection the superficial (S) and the deep veins (D) are superimposed. (B) Rotation of the limb separates the two sets of veins, the deep veins remaining close to the bones.

TECHNIQUES 35

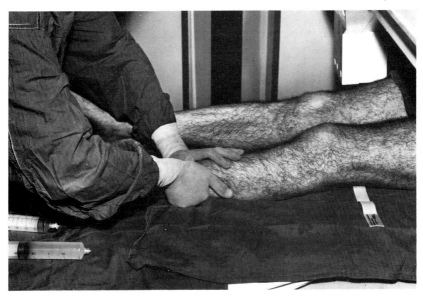

Fig. 3.13 The filling of the common femoral and iliac veins can be improved by compressing the calf and releasing the above knee tourniquet.

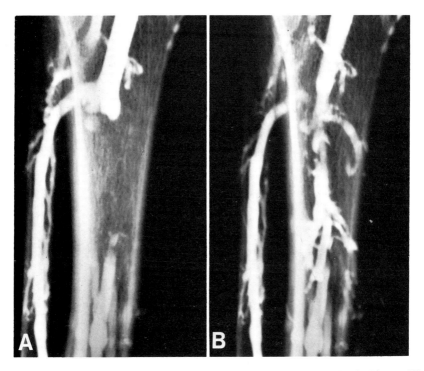

Fig. 3.14 (A) An unfilled segment of the peroneal veins which might be confused with non-filling due to thrombus. (B) It fills out fully with calf compression.

When the common femoral and the common iliac veins are filled with contrast, a Valsalva manoeuvre by the patient will show the profunda femoris and internal iliac veins are far as competent valves allow.

In order to keep the vein open while the films are being checked intermittent injections of physiological saline are made.

At the completion of the examination the veins are cleared of contrast medium by flushing with physiological saline and by exercising the calf muscles.

MODIFICATIONS OF THE STANDARD TECHNIQUE

These may be required in some situations when more detailed information is required of parts of the venous system.

The bolus technique to show the iliac veins and inferior vena cava

This technique has already been referred to as a method of filling the large proximal veins as part of the standard technique for ascending phlebography. If an adequate demonstration of the iliac veins and inferior vena cava has not been obtained by this method it is supplemented as follows: The screening explorator,

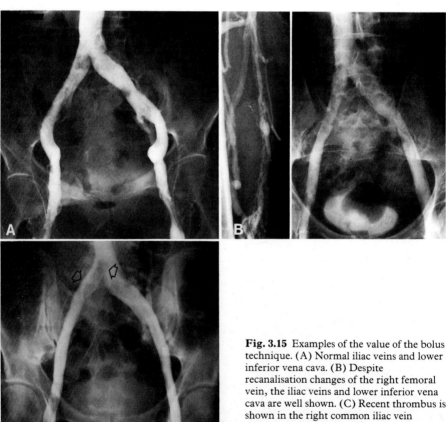

Fig. 3.15 Examples of the value of the bolus technique. (A) Normal iliac veins and lower inferior vena cava. (B) Despite recanalisation changes of the right femoral vein, the iliac veins and lower inferior vena cava are well shown. (C) Recent thrombus is shown in the right common iliac vein extending into the inferior vena cava (arrows). Separate per-femoral iliac phlebography was unnecessary. There is thrombus in both common femoral veins.

loaded with a 35 cm² film, is centred over the upper part of the sacrum and locked into position. A tourniquet is applied above each knee and 50 ml of meglumine iothalamate '280' injected simultaneously into each leg. The table is tilted slightly head downwards, the tourniquets are released and firm pressure applied to both calves by an assistant while the exposure is made (Fig. 3.15). This technique produces an adequate demonstration of the iliac veins and lower inferior vena cava in 95 per cent of patients. In the remaining patients separate phlebography by direct femoral injection or intraosseous injection is required.[24]

Examination of the foot veins

Foot phlebography is rarely required alone but a view of the foot should always be taken when ascending phlebography is being performed to investigate venous thrombosis as these veins sometimes contain thrombus.

In order to demonstrate as many of the foot veins as possible the injection of contrast medium should be made as far distally as practicable and a tight tourniquet used to fully distend the veins of the foot. The routine film is an oblique projection with the foot externally rotated and with the toe pointing downwards. Further projections may be required if the findings are equivocal (Fig. 3.16).

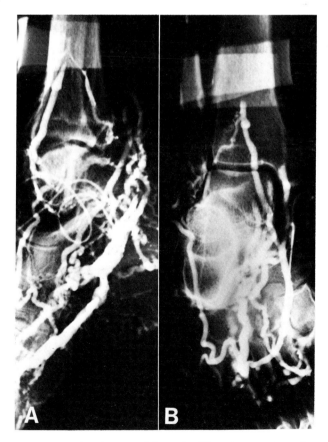

Fig. 3.16 Foot phlebography showing normal foot veins. There is a tourniquet above the ankle and the injection is made as distally as possible. (A) Standard oblique projection. (B) Straight projection.

VARIATIONS OF ASCENDING PHLEBOGRAPHY

While the principle of ascending phlebography has remained unchanged for many years, there are differences of opinion as to specific details of technique. The standard procedure described above has proved satisfactory in the author's hands and is easy to carry out without specialised equipment. There is rarely justification for a phlebographic X-ray room equipped solely for this purpose.

Some phlebographers still prefer the use of an overcouch tube.[9,16,30] A long cassette may be used so that the veins of one or both legs are shown on a single film.[9] An ankle tourniquet is not employed by some authors who rely on a steep table tilt to fill the deep venous system.[31] Others use no table tilt, preferring to examine the patient in the horizontal position.[29,30] The leg may be bandaged to clear blood from the soleal veins before the examination – a practice which is claimed to produce better filling of the muscle veins.[30]

Another variant is *tilt phlebography* where the patient is tilted feet down immediately after the injection in order to temporarily arrest the upward flow of blood and reverse the hydrostatic gradient so as to sharply identify venous valves and their competence.[11,28]

Dynamic *exercise phlebography* is carried out with the patient in a foot down position combined with leg muscle exercise[1,9] (Fig. 3.17). In this way the deep veins are filled by calf muscle pump action and the competence of communicating veins evaluated especially if cine fluoroscopy is used. Dynamic exercise phlebography may be combined with venous pressure measurements to study the musculovenous pump action.[2]

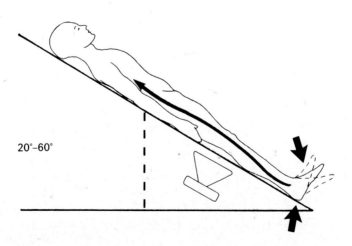

Exercise phlebography

Fig. 3.17 The principle of exercising ascending phlebography. (Redrawn after Ludbrook, 1972)[27].

PERCUTANEOUS ILIO-CAVAL PHLEBOGRAPHY

This method is used to demonstrate the iliac veins and inferior vena cava when the bolus technique has proved inadequate.

The patient lies supine on a fluoroscopy table equipped with a serial film changer. A control film is taken of the patient's pelvis, the field including the ischial tuberosities and the third lumbar vertebral body.

Both common femoral veins are punctured percutaneously using a needle or cannula. A trocar and cannula which can be threaded a short distance into the femoral vein to prevent displacement when moving the patient or during injection is preferred. A metal carotid cannula[32] (Fig. 3.18) or in thinner patients a disposable Potts-Cournand needle* are ideal for this purpose.

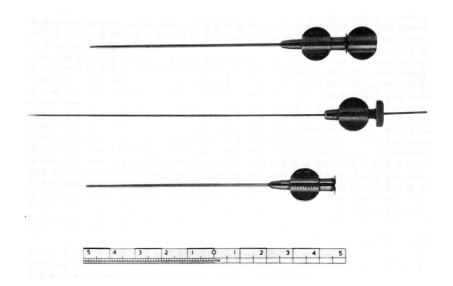

Fig. 3.18 Percutaneous femoral cannula. This one, designed for carotid arteriography, can be threaded into the femoral vein to prevent dislodgement using the short guide wire (middle). This is most useful in obese patients.[32] Otherwise the disposable Potts-Cournand needle (18 gauge) is convenient. It also can be threaded a short distance into the vein using a blunt metal obturator.

The common femoral vein lies medial to the femoral artery at the groin and thus provides a landmark for venepuncture. With a finger on the artery a stab is made just medial to it with the cannula. The trocar is then removed and the cannula is connected by an extension tube to a saline filled syringe. Gentle suction is applied to the syringe as the cannula is withdrawn. Venous blood enters the system with a successful puncture and the cannula can then be threaded further into the vein with a guide wire or blunt obturator. The puncture is facilitated if the patient is asked to perform the Valsalva manoeuvre to distend the vein. The examination is normally carried out using local anaesthesia but occasionally in nervous patients general anaesthesia is required.

*Becton-Dickinson, Rutherford, New Jersey

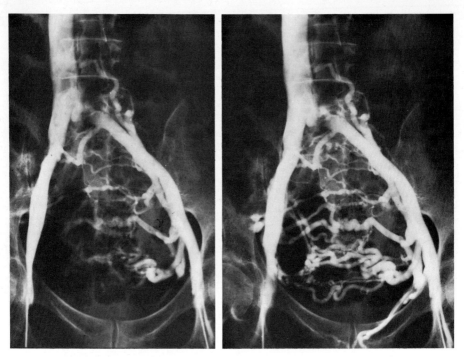

Fig. 3.19 Sequential films showing extensive collateral veins filling from left to right because of the left common iliac vein stenosis.

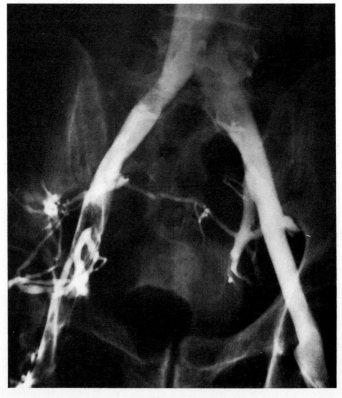

Fig. 3.20 Combined intraosseous (right) and perfemoral (left) iliocaval phlebogram. There is recent thrombus in the right common femoral and external iliac veins.

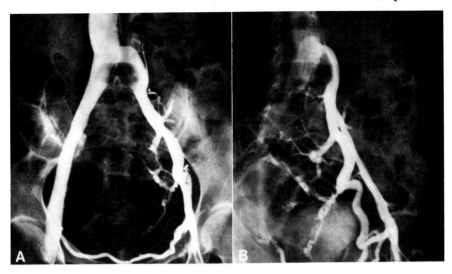

Fig. 3.21 (A) Iliocaval phlebogram in a straight projection. There is a suggestion of an iliac compression defect at the junction of the left common iliac vein with the inferior vena cava. (B) An oblique projection shows that this is a stenosis.

After test injections to check the position of the cannulae 50 ml of contrast is injected by hand simultaneously into both femoral veins and 10 films exposed at one per second using a serial film changer. The injection and filming sequence are started at the same time so that early films are obtained to show the direction of flow in any collaterals. The filming sequence not only demonstrates the venous anatomy but also allows for the delayed flow through collaterals in venous obstruction (Fig. 3.19).

A Valsalva manoeuvre during the injection will fill the internal iliac veins as far as competent valves permit. If one femoral vein is occluded or cannot be punctured a combined examination using a perfemoral injection on one side and an intraosseous injection (described later) on the other can be employed (Fig. 3.20).

Usually a straight projection is all that is required but this may be supplemented by oblique views if necessary (Fig. 3.21).

To show the inferior vena cava the procedure is similar to that described for the iliac veins except that the centering is at a higher level to include the inferior vena cava. As a rule the straight projection is adequate but this can be supplemented by oblique or lateral projections if necessary, or a simultaneous biplane series can be carried out.

TRANSCARDIAC INFERIOR VENA CAVOGRAPHY

If the lower inferior vena cava is occluded, the proximal extent of the occlusion may be shown by injecting contrast from above through a catheter passed from the right basilic vein through the right atrium into the upper inferior vena cava. The catheter is then advanced to the site of obstruction and withdrawn slightly before an injection of 50 ml of contrast is made by hand or pressure injector. A series of films is taken during the injection.

INTRAOSSEOUS PHLEBOGRAPHY

Intraosseous injection of contrast medium for phlebography was first carried out by Drasnar[13] but it was popularised by Arnoldi[2] and others.[5,17,23] An exhaustive study of its many uses is contained in the monograph by Schobinger.[33] The main use of intraosseous phlebography is to demonstrate the iliac veins and inferior vena cava when the femoral veins are occluded or impossible to puncture. The procedure is best avoided in patients whose epiphyses have not yet fused as inadvertent extravasation around the epiphyses may impair growth.

As intraosseous phlebography is painful it is carried out under general anaesthesia.

To facilitate penetration of bone, Lea Thomas[20] has devised a cannula with a triple facetted drill tip* (Fig. 3.22). The greater trochanter is identified by palpation while the leg is rotated internally and externally by an assistant. The cannula is inserted into the centre of the trochanter where the bone is thinnest (Fig. 3.23) with a firm 'screwing' action taking care to keep the trocar within the cannula, (Fig. 3.24), and its final position verified by test injection under screen control. When correctly positioned, contrast will be seen to enter the marrow cavity and then pass freely into the draining veins.

Fifty ml of contrast medium is injected simultaneously on each side using a pressure pump at 50 p.s.i. A series of 10 films of the pelvis is taken at the rate of 1 per second using a serial changer and starting the series at the beginning of the injection.

*V J Millard 36 Highgate Hill, London N19

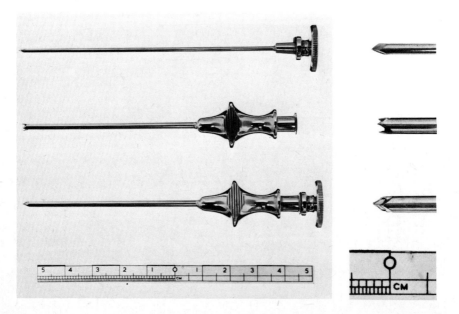

Fig. 3.22 The Lea Thomas 3-facetted trocar and cannula for intraosseous injections. The tip is shown in close-up

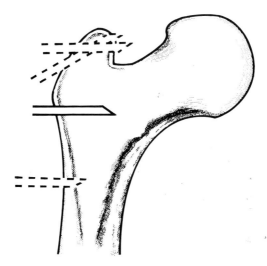

Fig. 3.23 Diagram of pertrochanteric intraosseous phlebography sites. The correct site of injection is the middle of the greater trochanter where the cortex is thinnest (continuous line). The cannulae with interrupted outlines are incorrectly placed. The position of the cannula should always be checked by a test injection under fluoroscopic control (Redrawn after Schobinger, 1960).

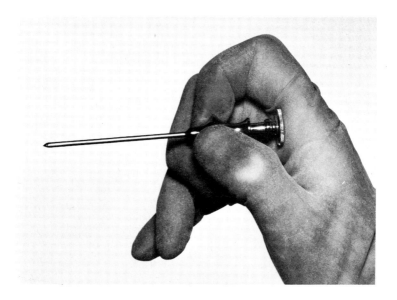

Fig. 3.24 The trocar is held tightly against the palm so that it cannot be displaced from inside the cannula as it penetrates the cortex.

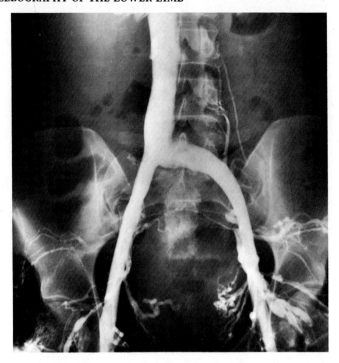

Fig. 3.25 A normal bilateral intraosseous phlebogram showing the iliac veins and lower inferior vena cava.

For a bilateral examination (Fig. 3.25) a separate injector should be used on each side because a single injection though a 'Y' connector is unsatisfactory as the contrast medium tends to pass into the side which presents the least resistance. Because of the slow flow of contrast medium from the bone marrow into the veins

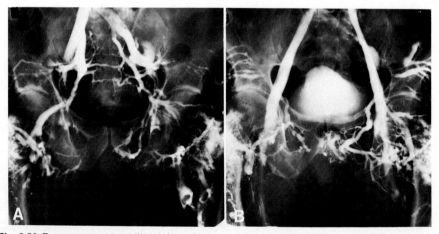

Fig. 3.26 Because contrast medium is hyperbaric compared with blood, it gravitates into the most dependent veins and parts of veins. This phenomena is particularly noticeable during intraosseous phlebography because a bolus of contrast cannot be delivered. (A) In the supine position the internal iliac venous systems are preferentially shown. (B) In the prone position the external and common iliac veins are better demonstrated.

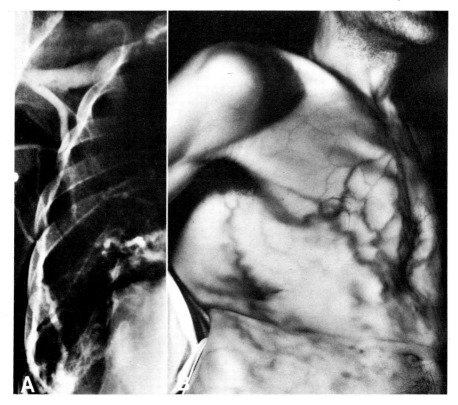

Fig. 3.27 (A) The lateral thoracic vein is shown by a right pertrochanteric intraosseous phlebogram in a patient with total inferior vena caval obstruction. (B) An infra red photograph of the patient showing superficial collaterals.

when the intraosseous method is used, the layering effect of contrast is more pronounced than with direct venous injection. For this reason a better display of the external and common iliac veins is obtained if the examination is carried out with the patient in the prone position. On the other hand if a full demonstration of the internal iliac veins and their tributaries is required the examination should be carried out in the supine position[22] (Fig. 3.26). When pertrochanteric intraosseous phlebography is used to show the inferior vena cava, bilateral injections are made centering higher and the sequence of films is started a few seconds later, about midway through the injection. In the presence of inferior vena caval obstruction centering over the superior vena cava may be needed to show collateral venous pathways (Fig. 3.27).

Intraossous phlebography can also be used to demonstrate veins in the lower legs when the percutaneous method is impossible or contra-indicated. Examples are severe oedema or where ankle ulceration prevents the application of an ankle tourniquet (Fig. 3.28). A small intraosseous cannula or bone marrow needle is introduced into the os calcis or into the medial or lateral malleolus. The medial malleolus is preferred as the marrow cavity is larger (Fig. 3.29). Intraosseous phlebography from the foot or ankle is probably the most accurate method of

46 PHLEBOGRAPHY OF THE LOWER LIMB

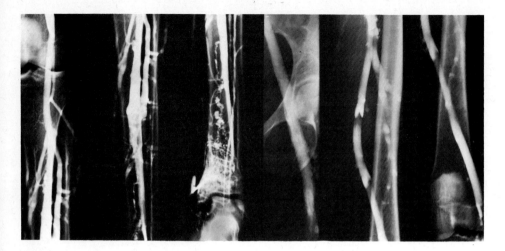

Fig. 3.28 A normal ascending phlebogram from an intraosseous medial malleolus injection. A little extravasation of contrast at the cannula site is present. Percuaneous venous injection was impossible because of oedema.

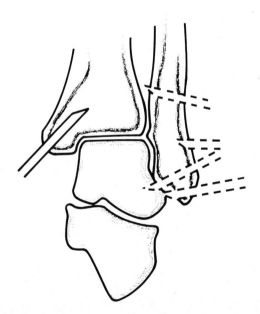

Fig. 3.29 Diagram of intraosseous injection sites at the ankle. The medial malleolus is the best site for intraosseous injection at the ankle as the bone marrow cavity is larger (continuous line cannula). The fibula cavity is small and may be missed by the cannula. The os calcis is an easy site to puncture but the patient usually complains of a painful heel on walking afterwards (Redrawn after Schobinger, 1960).

demonstrating incompetent communicating veins as the contrast medium tends to pass preferentially into the deep venous system by this route (Ch. 8).

Other sites which may be injected include the tibial tuberosity to show the femoral veins (Fig. 3.30) and the iliac crest to show the common iliac veins and inferior vena cava, and the pubic crest for the demonstration of the internal iliac system (Fig. 3.31).

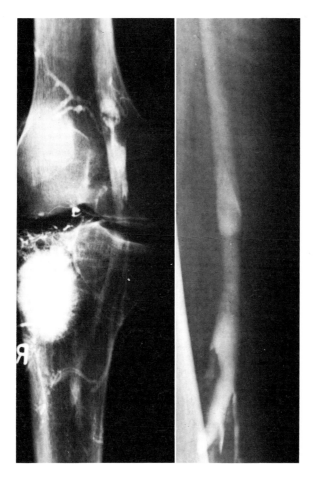

Fig. 3.30 A tibial tuberosity intraosseous ascending phlebogram. There is a little recent thrombus in the popliteal vein. The patient was a drug addict and the superficial veins were thrombosed.

An unusual application of intraosseous phlebography is to demonstrate deep venous thrombosis in the cadaver.[10]

The intraosseous technique is not often used these days with the introduction of selective catheterisation techniques but it remains a useful method in selected patients and situations.[33]

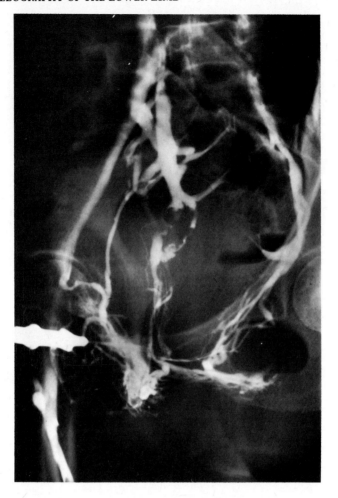

Fig. 3.31 A bone marrow aspiration needle in the right pubic bone used to demonstrate the venous drainage of vulval varices (see Ch. 11).

DESCENDING PHLEBOGRAPHY

The examination is carried out with the patient in the erect or semi-erect position (Fig. 3.32). The femoral vein is identified in the groin as described for percutaneous iliofemoral phlebography. The correct positioning of the cannula in the vein is confirmed with a small test injection under television control. Fifteen ml aliquots of contrast medium are then injected slowly by hand, the injection being made under fluoroscopic control and the contrast being allowed to descend into the leg with the patient standing on the foot rest of a fluoroscopy table. The events are recorded either on cut film or on ciné film. A Valsalva manoeuvre may be used to assist retrograde flow of contrast medium (Fig. 3.33) but the upright posture is usually adequate (Fig. 3.34). When the technique was first described it was thought to distinguish between competent and incompetent valves.[4, 6, 26]

TECHNIQUES 49

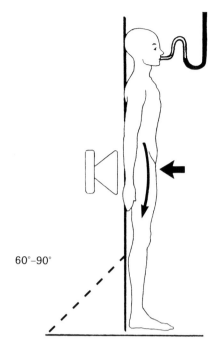

Descending phlebography

Fig. 3.32 Principle of descending phlebography. A steep table tilt or the erect position alone, or combined with a Valsalva manoeuvre is used. (Redrawn after Ludbrook 1972)[27]

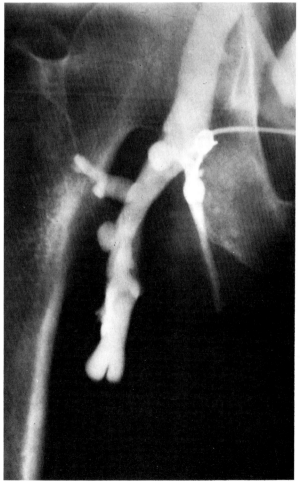

Fig. 3.33 Descending phlebogram with a Valsalva manoeuvre. The valves are distended and appear competent.

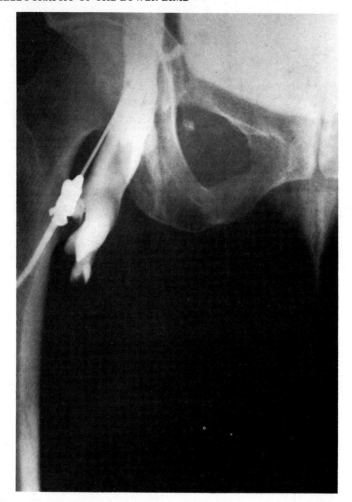

Fig. 3.34 Descending phlebogram in the upright position without a Valsalva manoeuvre. Valve cusps are clearly defined and appear competent.

Further experience of valve function[8,15] has shown that contrast medium can flow past normal valves (Fig. 3.35). The degree of retrograde flow in normal veins in the upright position depends partly on the specific gravity of the contrast medium which is greater than blood and partly on the force of injection. A forceful retrograde flow such as that produced by a Valsalva manoeuvre will close the valves so that little contrast descends into the leg, whereas a slow injection will flow past normal valves.

There has been renewed interest in this technique since the work of Kistner[19] and the use of transvenous repair of incompetent femoral valves. Kistner believes that a competent valve will hold up contrast medium in a patient breathing normally in a 60° head up position whereas with severe valvular incompetence there is leakage of most of the contrast medium downwards through the valves and little, or no, central flow.

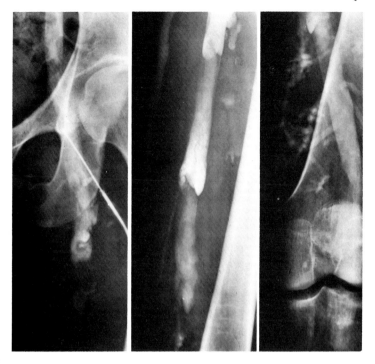

Fig. 3.35 A descending phlebogram showing retrograde flow past femoral valves. The valve cusps are present but irregular and appear to be incompetent.

The author and his surgical colleagues find it difficult to judge the competence of femoral valves using this technique.

RETROGRADE PHLEBOGRAPHY

Retrograde phlebography is a modification of descending phlebography and is often carried out following pulmonary angiography. A catheter is passed from an arm vein through the right atrium, down the inferior vena cava into the femoral vein. There are no valves in either the inferior vena cava or in the iliac veins and the catheter can be readily passed distally as far as the junction of the external iliac and common femoral veins. The catheter can also be positioned in either internal iliac vein. Occasionally some difficulty may be encountered in avoiding the hepatic and right renal veins. As the catheter (No. 8 or 9 N.I.H.) is advanced under fluoroscopic control, 10 to 15 mls of contrast is injected by hand at various levels and films taken using an undercouch tube. In this way the whole of the inferior vena cava, the common and external iliac veins can be clearly demonstrated. Injection of the internal iliac veins also demonstrates the pre-sacral venous plexus (Dow, 1973).

A similar principle is employed in the technique of iliac cross-over phlebography. The femoral vein is catheterised using the Seldinger technique employing a curved arterial catheter. The catheter is then advanced under fluoroscopic control and positioned in the opposite common iliac vein. Small injections of contrast are made as the catheter is advanced until the proximal limit of the occlusion is defined. A hand injection of 20 mls of contrast is then made while a series of films is exposed. The main indication is to show the upper extent of a thrombotic occlusion of the iliac or femoral vein. While it does not give as much information as intraosseous phlebography it is simpler to carry out and requires only local anaesthetic.

REFERENCES

1. Almen T, Nylander G 1962 Serial phlebography of the normal lower limb during muscular contraction and relaxation. Acta Radiol 57: 264
2. Arnold C C, Bauer G 1960 Intraosseous phlebography. Angiology 11: 44
3. Arnoldi C C, Greitz T, Linderholm H, 1966 Variations in cross sectional area and pressure in the veins of the normal human leg during rhythmic muscular exercise. Acta chir scand 132: 507
4. Bauer G A 1948 The aetiology of leg ulcers and their treatment by resection of the popliteal vein. J Int Chir 89: 37
5. Begg A C 1954 Intraosseous phlebography of the lower limb and pelvis. Brit J Radiol 23: 318
6. Boyce W H, Datar J H, Vest S A 1953 A new technique of venography of the lower extremities with Urokon. Surg Gynec Obstet 96: 471
7. Buist T A S 1975 Phlebography in pulmonary embolism. In: (eds) Vaughan Ruckley C and MacIntyre I M C Venous thrombosis and embolic disease Churchill Livingstone, Edinburgh
8. Cockett F B 1953 The practical uses of venography. Brit J Radiol 26: 339
9. De Weese J A, Rogoff S M 1959 Functional ascending phlebography of the lower extremity by serial long film technique. Amer J Roentgenol 81: 841
10. Diener L 1971 Intraosseous phlebography of the lower limb; Post mortem investigation of thrombotic venous disease. Acta Radiol Suppl 304
11. Dohn K 1958 Tilt phlebography; retrograde phlebography by ascending injection. Acta Radiol 50: 293
12. Dow J D 1973 Retrograde phlebography in major pulmonary embolism. Lancet 2: 407
13. Drasnar U 1946 Intraspongiose Dauertropfinfusion. Schweiz med Wschr 76: 36
14. Gothlin J 1972 The comparative frequency of extra-vasal injection at phlebography with a steel and plastic cannula. Clin Radiol 23: 183
15. Gryspeerdt G L 1953 Venography of the lower limb. Brit J Radiol 26: 329
16. Haeger K, Nylander G 1967 Acute phlebography. Triangle 8: 18
17. Halliday J P 1967 Intraosseous phlebography of the lower limb. Brit J Surg 54: 248
18. Halliday J P 1968 Phlebography of the lower limb. Brit J Surg Engl 55: 226
19. Kistner R L 1975 Surgical repair of the incompetent femoral vein valve. Arch Surg 110: 1336
20. Lea Thomas M 1969 An improved intraosseous phlebography cannula. Br J Radiol 42: 395
21. Lea Thomas M 1972 Phlebography. Arch Surg 104: 145
22. Lea Thomas M, Browse N L 1972 Internal iliac vein thrombosis. Acta Radiol Diag 12: 660
23. Lea Thomas M, Fletcher E W L 1967 The techniques of pelvic phlebography. Clin Radiol 18: 399
24. Lea Thomas M, MacDonald L 1977 The accuracy of bolus ascending phlebography in demonstrating the ilio-femoral segment. Clin Radiol 28: 165
25. Lea Thomas M, O'Dwyer J A 1978 A phlebographic study of the incidence and significance of venous thrombosis of the foot. Am J Roentgenol 130: 751
26. Lockhart-Mummery H E, Hiller-Smith J 1951 Varicose ulcer. A study of the deep veins with special reference to retrograde venography. Brit J Surg 38: 384
27. Ludbrook J 1972 The analysis of the venous system. Hans Huber Publishers, Bern
28. Mathiesen F R 1958 Tilt phlebography of normal legs. Acta Radiol 50: 493

29. Nicolaides A N, Kakkar V V, Field E S, Renney J T G 1971 The origin of deep vein thrombosis; a venographic study. Brit J Radiol 44: 653
30. O'Dell C W, Coel M N 1976 Continuous infusion supine phlebography. Journal de L' Association Canadienne de Radiologistes 27: 186
31. Rabinov K, Paulin S 1972 Roentgen diagnosis of venous thrombosis in the leg. Arch Surg 104: 134
32. Sheldon P 1964 Percutaneous cannulation of the carotid artery. Brit J Radiol 37: 526
33. Schobinger R A 1960 Intraosseous venography. Grune and Stratton, New York.

4

Complications

INTRODUCTION

The conventionally used contrast media in phlebography are tri-iodinated benzoic acid derivatives which are mono-acid monomers (Fig. 4.1). These include the two media meglumine iothalamate '280' and sodium iothalamate '420' used by the author.

These media are highly hyperosmolar, 5 to 7 times that of plasma. This hyperosmolality has generalised effects producing vasodilatation, fluid shifts resulting in crenation of red blood corpuscles, expansion of the plasma volume, rouleaux formation, and of particular importance to the phlebographer, endothelial damage.

It is probable that this endothelial injury together with tissue damage when extravasation occurs, accounts for many of the complications of phlebography.

The osmolality of contrast media can be reduced in several ways. The most obvious one is to dilute the contrast medium but this has limitations because the iodine content may become too low to produce diagnostic phlebograms. The most

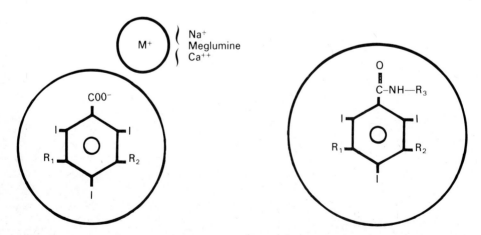

Fig. 4.1 General formula of a conventional contrast medium. This is a monoacid monomer of high osmolality – 1.5 osmols/kg H_2O, compared with 0.3 osmols/kg H_2O for plasma.

Fig. 4.2 General formula of a nonionic medium. This has a low osmolality. Examples are metrizamide '280' (0.47 osmols/kg H_2O) and iopamidol '280' (0.47 osmols/kg H_2O).

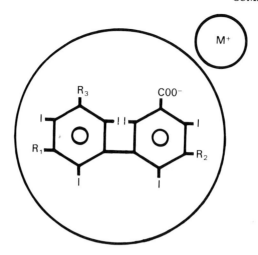

Fig. 4.3 The formula of another low osmolality contrast medium. This is a monoacid dimer (ioxaglate) with an osmolality of 0.49 osmols/kg H_2O.

logical way is to maintain the iodine content at diagnostic levels and lower the osmolality either by making the molecule non-dissociable in solution, producing a non-ionic medium (Fig. 4.2), or by increasing the molecular weight by attaching a second benzene ring producing a dimer (Fig. 4.3).

The osmolality of the non-ionic contrast media metrizamide and iopamidol each containing 280 mg/ml of iodine are both 0.47 osmols/kg H_2O compared with meglumine iothalamate '280' which has an osmolality of 1.5 osmols/kg H_2O. The osmolality of plasma is 0.3 osmols/kg H_2O.

GENERALISED SYSTEMIC AND IDIOSYNCRATIC REACTIONS

These complications are not specific to phlebography but are common to all situations where contrast media are used.

Minor reactions such as a feeling of warmth, a metallic taste in the mouth are frequent. Sneezing, nausea, retching, vomiting and mild urticaria occur in about 8 per cent of patients,[32] usually require no treatment and pass off rapidly. More severe reactions such as bronchospasm or glottic oedema require immediate specific treatment with drugs such as bronchodilators, intravenous antihistamines hydrocortisone or adrenaline. Loss of consciousness, cardiac arrythmias and arrest, pulmonary oedema, and myocardial infarction require instant intensive therapy to prevent death.

Severe reactions are unpredictable and inconstant. The response to a test injection is unreliable and may itself give rise to a reaction or even death and so test injections are no longer used.

The mortality rate related to the intravenous injection of contrast medium is of the order of 1 in 40 000 examinations.[3] Appropriate drugs and life saving equipment must therefore always be at hand.

LOCAL COMPLICATIONS
Pain
Pain is the commonest, albeit minor, complication of phlebography. Its exact cause is not fully understood but it is considered to be related to the sodium content and the hyperosmolality of the media.[11,14] Pain is more severe in patients with obstructed venous systems due to venous thrombosis or other causes. Pain can be reduced in severity by using a sodium free contrast medium and by diluting it. Schmitt[36] recommends a sodium free contrast medium such as meglumine diatrizoate 65% diluted to 40% and Bettman and Paulin[6] suggests diluting Renografin 60% to three quarters concentration with 5% dextrose saline. The latter found that the pain associated with ascending phlebography was reduced from 59 per cent to 30 per cent in this way while maintaining the diagnostic quality of the phlebograms. The author does not dilute the contrast medium because the iodine content of the diluted medium, although sufficient in most cases for phlebography of the legs, is not sufficiently dense to clearly demonstrate the veins of the thigh and pelvis. Meglumine iothalamate 60% (Conray '280') is used for ascending phlebography of the legs. This is virtually sodium free (1.2 meq/l) and has a low viscosity (4 cp) allowing it to be more easily injected through needles.

For the larger proximal veins and inferior vena cava, sodium iothalamate 70% (Conray '420') is usually employed to provide fully diagnostic phlebograms. This sodium containing medium is used because of the relatively low viscosity so that it can be easily injected through cannulae into the femoral veins. Pain is much less frequent when contrast is injected into the femoral veins presumably due to their larger calibre with consequent dilution.

As suggested above, it is probable that in the future the low osmolality media will be used routinely for phlebography. In a recent study[25] meglumine iothalamate '280' was injected into one leg and metrizamide of equivalent strength injected into the other. This double-blind prospective trial in which the patient acted as his own control indicated that 68 per cent of patients complained of pain in the leg injected with meglumine iothalamate and only 15 per cent with metrizamide.

Tourniquets used in the authors standard technique sometimes produce a bursting sensation in the foot and calf which is not intolerable and passes off as soon as they are removed.

Severe pain complained of by patients should be taken seriously as it may indicate that extravasation into the tissues has occurred. This may be the result of too tight a tourniquet producing bursting of distal veins or from displacement of the needle. The site of the pain should always be examined immediately and fluoroscopy carried out to exclude extravasation. When examining for incompetent communicating veins the ankle tourniquet must be tight enough to occlude the superficial venous system as described in Chapter 8, but if this cannot be achieved with a reasonably tight tourniquet the examination should be abandoned and an intra-osseous technique used instead.

Thrombosis
Thrombotic complications have been reported since the early days of phlebography but have been considered rare and of little clinical significance.[14,16,21,34,40]

It is an established fact that hyperosmolar solutions injure the vein walls and predispose to thrombosis.[2, 28, 29, 30, 35, 42] It has been suggested that the recently introduced low osmolality media do not cause thrombosis.[1] This observation has been largely confirmed by the study mentioned above.[39] Only 7 per cent of patients examined with metrizamide developed scintigraphic evidence of venous thrombosis whereas 28.6 per cent of patients had fibrinogen uptake diagnosed thrombosis when meglumine iothalamate was used.

It would thus be reasonable to assume that the delayed leg pain experienced by a number of patients examined using a conventional medium may be due to episodes of venous thrombosis. In the writers experience[22] about 30 per cent of patients complain of calf pain within four days of a phlebogram. In Bettman and Paulin's experience[6] clinical signs and symptoms of venous thrombosis following phlebography were reduced from 24 per cent to 7.5 per cent by diluting the contrast medium.

Leaving aside these new low osmolality media which will shortly be generally available, it must be accepted that a certain percentage of patients undergoing ascending phlebography will develop venous thrombosis. It is important to minimise the risk of damage to the endothelium which predisposes to thrombosis by carrying out the examination as rapidly as possible, clearing the contrast medium from the veins with physiological saline at the end of the examination, and assisting the clearance from the veins by local massage, elevation of the limb, passive movement and early ambulation. Since introducing these measures the incidence of clinically diagnosed postphlebographic thrombosis has been reduced to about 10 per cent, the thrombosis usually occurring in the vein into which the contrast has been injected.

The routine use of heparin before and after the examination in doses of 2500 to 5000 units to minimise the risk of thrombus has been advocated.[7, 17, 28] The author however reserves this for patients with a high thrombotic tendency such as those with carcinomatosis. The individual response to heparin is very variable and the dose employed may be insufficient to prevent thrombosis or conversely so great that there is a risk of haemorrhage.

Generally, it is wiser not to carry out phlebography on anticoagulated patients because of the risk of haemorrhage at the venepuncture or intraosseous sites. However, in emergency situations, anticoagulation is not a contraindication for phlebography but additional pressure at the puncture site should be used to promote haemostasis.

Local extravasation
If contrast medium extravasates into the tissues it causes a chemical cellulitis. Sometimes this may progress to ulceration and soft tissue necrosis (Fig. 4.4). Very rarely gangrene occurs (Fig. 4.5). This maybe the result of venous thrombosis or an extreme degree of perivascular inflammatory tissue reaction.

Though rare, these complications may have severe consequences, occassionally requiring skin grafting and even amputation. Soft tissue necrosis is more common in patients with arterial or venous insufficiency or when the deep venous system is absent or occluded[5, 10, 19] (see Fig. 4.5).

Extreme care should be taken over the venepuncture, and the position of the

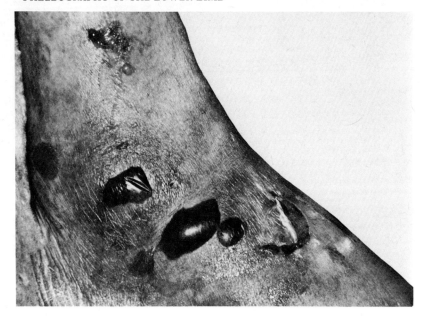

Fig. 4.4 Superficial skin necrosis and blistering following extravasation of contrast medium.

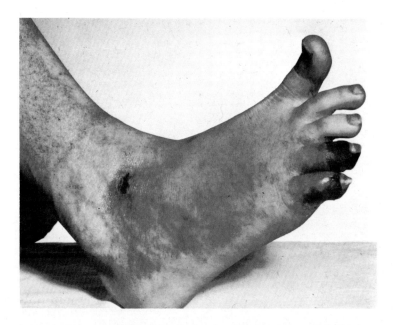

Fig. 4.5 Gangrene of the fourth and fifth toes two days following an ascending phlebogram. The patient has Klippel-Trenaunay syndrome with absent deep veins of the calf preventing clearing of the contrast from the foot. The site of a cut down for the phlebogram can be seen in front of the lateral malleolus.

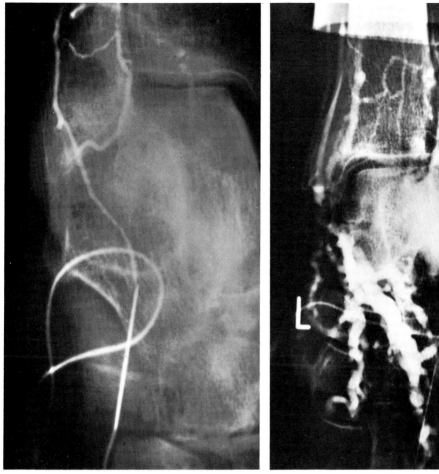

Fig. 4.6 **Fig. 4.7**

Fig. 4.6 Correct position of the needle for phlebography showing on a test injection of contrast medium that there is no extravasation.

Fig. 4.7 Extravasation has occurred below the lateral malleolus from an unsuccessful venepuncture. The injection was stopped, but the needle left in position to minimise leakage at this puncture site during injection from a further venepuncture.

needle checked by injecting physiological saline and then a test injection of contrast medium under fluoroscopic control before proceeding (Fig. 4.6). Throughout the examination the site of injection should be inspected frequently for swelling and examined by fluoroscopy for evidence of extravasation. If this is found to have occurred the injection should be stopped and a new venepuncture made at a different site leaving the first needle *in situ* to minimise leakage from the previous puncture (Fig. 4.7). If the extravasation is at all severe (Fig. 4.8) the examination should be stopped.

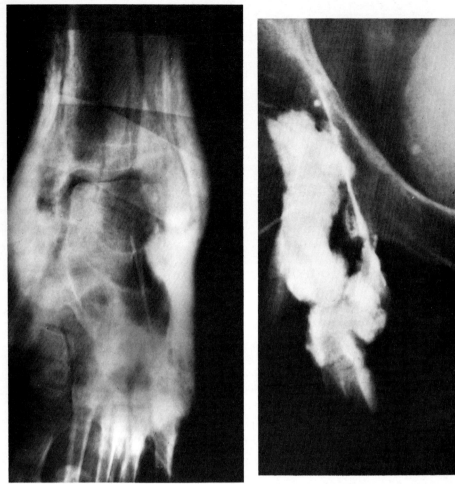

Fig. 4.8 Gross extravasation of contrast medium probably due to too tight an ankle tourniquet causing rupture of distal fragile veins.

Fig. 4.9 Extravasation of contrast medium around the femoral vein which shows as a relative translucency. This is not usually a serious complication as the contrast medium is rapidly absorbed but a stinging sensation occurs for a few minutes.

Injection of hyaluronidase has been advocated when extravasation occurs.[9] The combination of this drug and contrast medium increases the risk of tissue damage.[31] In the author's view the best method of dispersing the contrast medium is by gentle local massage and diluting the contrast medium in the tissues by injecting physiological saline.

Some consider[9] that there is less risk of extravasation if a plastic cannula is used rather than a needle, but a needle is sharper and venepuncture is easier.

Extravasation into the tissues around the femoral vein (Fig. 4.9) does not appear to be of such serious consequence because the contrast medium is readily absorbed from the surrounding bulk of tissues although the patient will complain of a slight local stinging pain for a few minutes.

COMPLICATIONS OF INTRAOSSEOUS PHLEBOGRAPHY

This method involves the injection of contrast medium through a bone cannula[18] into the marrow cavity, as discussed in Chapter 3. It is a useful technique for the demonstration of the venous circulation when conventional intravenous methods are impossible.

Whatever injection site is chosen care must be taken to ensure that the end of the cannula lies within the bone marrow cavity. This is usually indicated when blood oozes from a cannula when withdrawing the trocar, but positive aspiration is often required to confirm. The position of the cannula should be checked by a test injection under fluoroscopic control before proceeding with the examination. If the test injection shows that the cannula is incorrectly placed the first cannula should be left in position and another introduced (Fig. 4.10). Unless this is done contrast medium extravasates from the first puncture site. Incorrect positioning of the cannula results in extravasation into the soft tissues, injection beneath the periosteum or into a joint (Fig. 4.11).

These complications are not usually serious but cause the patient some pain for a few days. Haemarthrosis may be a troublesome complication resulting in pain on movement of the joint.

If the os calcis is used for injection, pain on walking for some time afterwards is quite common and for this reason this site should be avoided if possible.

Although intraosseous phlebography has been extensively used serious complications are rare.[20,37] Osteomyelitis at the site of bone puncture is clearly a possibility, particularly if the bone puncture is made close to ankle ulceration. A site well away from areas of ulceration such as the malleoli should therefore be chosen for injection. Sometimes the os calcis has to be used despite the risk of pain.

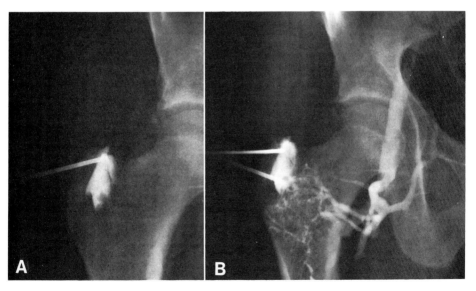

Fig. 4.10 (A) The cannula has been introduced subperiosteally. (B) A second cannula has been inserted leaving the first one in place to avoid leakage of contrast medium from the initial puncture site at the time of the main injection.

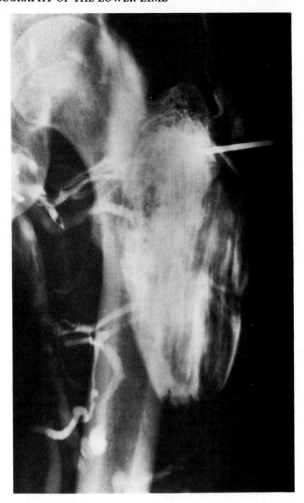

Fig. 4.11 Pertrochanteric intraosseous phlebogram with extravasation into the muscle planes.

Full aseptic precautions should be taken during intraosseous phlebography and preferably an X-ray room used which is reserved for special procedures.

Bone infarction has been recorded,[15] but this complication has not been encountered by the author in over a thousand examinations even though follow up films up to a year after the investigation have been taken in a number of patients. Pulmonary embolism has been recorded following pertrochanteric iliac phlebography[27] but this was considered more likely to be due to the general anaesthetic than the procedure itself.

The most serious, but fortunately rarest complication of intraosseous phlebography, is fat embolism which may be fatal.[23,41] It has been suggested by Schobinger[37] that pressure injections should not be used for intraosseous phlebography. However, to produce fully diagnostic phlebograms when examining large veins such as those in the pelvis, pressure injections are essential. Any patient developing neurological or pulmonary symptoms or signs 12 hours or more after intraosseous phlebography should be suspected of having fat embolism. Investigations should include a search for fat globules in the sputum, urine and blood. Treatment of the condition requires continuous oxygen and intravenous injection of low molecular weight dextran. Following a death from fat embolism[23] the blood of all patients undergoing intraosseous phlebography has been examined after the examination for fat globules. None has been found.

COMPLICATIONS OF ILIOCAVAL PHLEBOGRAPHY

Complications following iliocaval phlebography are extremely rare: no serious complications have occurred in over 800 examinations. The only minor complication which may occur using the percutaneous femoral technique is haematoma at the puncture site. This low morbidity is the experience of others.[11,26,33]

When a guide wire is used to thread a catheter up the femoral vein, perforation of the iliac vein or the inferior vena cava is a possibility. This occurrence is usually without sequelae[13] but occasionally haemorrhage into the abdomen or pelvis may have serious consequences. For this reason the author prefers to use a needle or short cannula for investigations of the iliac veins and the inferior vena cava unless selective catheterisation of a tributary of these veins is essential.

CARDIAC COMPLICATIONS

Patients with pulmonary hypertension

Phlebography is often requested in patients with pulmonary hypertension because this may be caused by repeated episodes of minor pulmonary embolism. Also since patients with cardiac lesions are more prone to venous thrombosis, patients with unrelated pulmonary hypertension may also require investigation.

Such patients are particularly liable to develop cardiac arrhythmias and arrest which are often very difficult to reverse. These result from a reduction in cardiac output due to the profound vasodilatation produced by contrast media. In such patients the pulmonary artery pressure should be continuously monitored throughout the examination. Cardiac complications are more likely to occur when the contrast is delivered rapidly as a bolus but the risk is present whenever contrast is injected. Vasodilatation can be counteracted by the use of a vasoconstrictor before injection of contrast medium. Phenylephrine 0.3 mg is given intravenously and repeated if necessary until the systolic blood pressure is about 20 mm Hg above its original level. Volume overload by the injection of 200 ml of plasma or plasma expander also helps to maintain the cardiac output. Even taking these precautions phlebography carries a significant morbidity in patients with pulmonary hypertension.

Air embolism

This is a theoretical complication of phlebography. To avoid it, syringes and connecting tubing should be free of air, filled either with physiological saline or contrast medium throughout the examination. This applies whether hand or pump injections are used. Should air embolism occur, the injection should be immediately stopped and the patient turned on his left side in the head-down position, so that air in the right ventricle floats away from the pulmonary outflow tract.

Pulmonary embolism

The author has not encountered a single case in which pulmonary embolism could be attributed directly to a phlebogram, even when recent thrombosis has been demonstrated. This is also the experience of others.[12,34] On the other hand there have been a few reports attributing pulmonary embolism to the examination.[40]

The technique[24] of calf compression to produce a bolus of contrast medium to show the iliac veins and inferior vena cava may be criticised as likely to dislodge nonadherent thrombus. Calf compression is however, a standard part of the Doppler technique[8] and the compression employed is probably no more than that produced by the patient's own muscular contractions or that used clinically to assess calf tenderness. The benefit of the bolus technique which often makes a separate iliac phlebogram unnecessary, outweighs this theoretical complication.

The routine use of the Valsalva manoeuvre to demonstrate the internal iliac and profunda veins from foot injections,[21] can also be criticised on the theoretical grounds that dilatation of the pelvic veins can loosen thrombus. The subsequent deep inspiration following the manoeuvre increases venous return and may suck thrombus into the lungs.[27] Again this theoretical risk has to be weighed against the advantage of avoiding separate iliac phlebography.

REFERENCES

1. Albrechtsson U, Olsson C G 1979 Thrombosis after phlebography: A comparison of two contrast media. Cardiovasc Radiol 2: 9
2. Almen T, Hartel M, Nylander G, Olivercrona N 1973 Effects of metrizamide on silver staining of the aortic endothelium. Acta Radiol Suppl 335: 233
3. Ansell G 1970 Adverse reactions to contrast agents: Scope of problem. Invest radiol 5: 374
4. Bartley O 1958 Venography in the diagnosis of pelvic tumours. Acta Radiol 49: 169
5. Berge T, Bergqvist D, Efsing H O, Hallböök T 1978 Local complications of ascending phlebography Clin Radiol 29: 691
6. Bettman M A, Paulin S 1977 Leg phlebography: The incidence, nature and modification of undesirable side effects. Radiology 122: 101
7. Corrigan T P, Fossard J, Spindler J, et al 1964 Phlebography in the management of pulmonary embolism. Br J Surg 61: 484
8. Evans D S, Cockett F B 1968 Diagnosis of deep venous thrombosis using an ultrasonic Doppler technique. Br Med J 2: 802
9. Gothlin J 1972 The comparative frequency of extravasal injection at phlebography with steel and plastic cannula. Clin Radiol 23: 183
10. Gothlin J, Hallböök T 1971 Extravasal injection of contrast medium at phlebography. Der Radiologe 4: 161
11. Gullmo A 1964 Periphere venen. Handbuch der Med Radiol Band X
12. Haegar K, Nylander G 1967 Acute phlebography. Triangle 8: 18
13. Hipoma F A 1969 In: Ferris E J et al (eds) Venography of the inferior vena cava and its branches. The Williams and Wilkins Co, Baltimore p 33
14. Homans J 1942 Thrombosis as a complication of venography. Jama 119: 136
15. Isherwood I 1972 In: Saxton H and Strickland B (eds) Practical procedures in diagnostic radiology. 2nd edition. H K Lewis and Co Ltd, London p 272
16. Kahr E 1953 Darstellung der Beckenvenen mittels transossarer Serienphlebographic. Fortschr Geb Roentgenstr Nuklearmed 78: 449

17. Kakkar V V 1972 The ^{125}I fibrinogen test and phlebography in the diagnosis of deep vein thrombosis. Millbank Mem Fund Q50 (Suppl 2): 206
18. Lea Thomas M 1969 An improved intraosseous phlebography cannula. Br J Radiol 42: 395
19. Lea Thomas M 1970 Gangrene following peripheral phlebography of the legs. Br J Radiol 43: 528
20. Lea Thomas M 1976 In: Ansell A (ed) Complications of diagnostic radiology. Blackwell, Oxford
21. Lea Thomas M 1972 Phlebography. Arch Surg 104: 145
22. Lea Thomas M, MacDonald L M 1978 Complications of phlebography of the leg. Brit Med J 2: 307
23. Lea Thomas M, Tighe J R 1973 Death from fat embolism as a complication of intraosseous phlebography. Lancet 2: 1415
24. Lea Thomas M, McAllister V, Tonge K 1971 The radiological appearances of deep vein thrombosis. Clin Radiol 22: 295
25. Lea Thomas M, Walters H L 1979 Metrizamide in ascending venography of the legs. Brit Med J 2: 1036
26. Lindblom A 1960 In: McLaren J W (ed) Modern trends in diagnostic radiology. Third series Butterworth, London p 111
27. Mahaffy R G, Mavor G E, Galloway J M D 1971 Iliofemoral phlebography in pulmonary embolism. Brit J Radiol 44: 172
28. May R 1965 The early X-ray diagnosis of thrombosis. Nuclear Energy 4: 120
29. May R 1977 Thrombophlebitis. Nach Phlebographie Vasa 6: 169
30. Mesereau W A, Robertson H R 1951 Observations on venous endothelial injury following the injection of various radiographic contrast medium in the rat. J Neurosurg 18: 289
31. McAllister W H, Palmer K 1971 The histological effects of four commonly used contrast media for excretory urography and an attempt to modify the response. Radiology 99: 511
32. Ochsner S F, Calonje A 1971 Reactions to intravenous iodides in urography. Surg Med J 64: 907
33. Ranniger K, Saldino R M 1968 Abdominal angiography. Current problems in surgery Yearbook Medical Publisher, Chicago p 28
34. Rabinov K, Paulin S 1972 Roentgen diagnosis of thrombosis in the leg. Arch Surg 104: 134
35. Ritchie W G M, Lynch P R, Stewart G J 1974 The effect of contrast media on normal and inflamed canine veins. A scanning and Transmission electron microscopic study. Invest Radiol 9: 444
36. Schmitt H E 1970 In: Kappert A (ed) New trends in venous diseases. Hans Huber Publishers, Bern
37. Schobinger R A 1960 Intraosseous phlebography. Grune and Stratton, New York and London
38. Ternberg J L, Butcher H R 1965 Evaluation of retrograde pelvicvenography. Arch Surg 91: 607
39. Walters H L, Clemenson J, Browse N L, Lea Thomas M 1980 ^{125}I fibrinogen uptake following phlebography of the leg. Radiology 135: 619
40. Werner H, Otto K 1962 Hazards and complications in Roentgenological venous diagnosis. Fortschr Roentgenstr 96: 655
41. Young A E, Lynn Edwards I, Irving D, Hamming C D 1973 Fat embolism after pertrochanteric venography. Br Med J 4: 592
42. Zinner G, Gottlob R 1959 Morphologic changes in vessel endothelia caused by contrast media. Angiology 10: 207

5

Artefacts

INTRODUCTION

Since the introduction of phlebography for the examination of the veins of the lower limb, there have been many improvements in the techniques which enable a more accurate evaluation of the venous anatomy or pathology.

These improvements, which have been discussed in more detail in Chapter 3, result from the use of larger volumes of contrast medium, the upright position to encourage mixing of blood and contrast medium, and tourniquets to produce better opacification of the deep venous system.

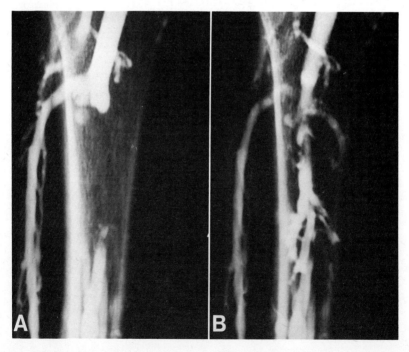

Fig. 5.1 (A) Non-filling of the peroneal veins suggesting occlusion by thrombus. (B) A later film following calf compression (bolus technique), fills out the veins indicating that the initial appearance was an artefact.

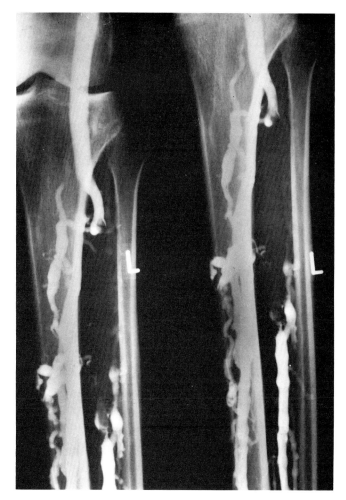

Fig. 5.2 There is a non-filled segment in the medial peroneal vein which is constant in two films taken with an interval between them. Careful inspection reveals a little thrombus at each end of the unfilled segment making the diagnosis of occlusion by acute thrombosis certain.

Difficulties of interpretation do however arise because of technical factors, the dynamics of venous flow, and because veins have thin walls and are easily deformed by neighbouring normal structures.

In this chapter various appearances which may mimic venous abnormalities such as acute venous thrombosis, postthrombotic sequelae and extrinsic deformity of the venous channels are described. Where possible, attention is drawn to ways in which such artefacts may be prevented or minimised, and the features which may enable them to be distinguished from actual disease processes are discussed.

APPEARANCES WHICH MIMIC VENOUS ABNORMALITIES

Non filling or under filling of veins

When a vein or a segment of vein remains unopacified and unchanged in appearance in two films taken with an interval between them the appearance suggests occlusion by a thrombus, but when such an appearance is seen only in a single film it is likely to be artefactual (Fig. 5.1). Careful inspection of the unfilled

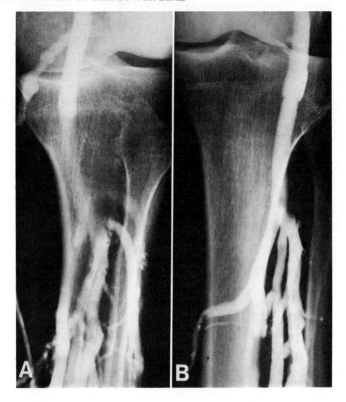

Fig. 5.3 (A) The distal part of the popliteal vein is under opacified. (B) A later film shows complete opacification indicating that the first appearance was artefactual.

segment may reveal a small amount of thrombus surrounded by contrast medium at either or both ends of the segment enabling a firm diagnosis of venous thrombosis to be made (Fig. 5.2).

Underfilling and delayed opacification of the lower part of the popliteal vein is common due to slight compression at this site by the heads of the gastrocnemius muscle and also because this part of the vein is the most dependent in the supine position, so that the hyperbaric contrast medium tends to gravitate to the posterior part of the vein. A similar appearance may be seen in the proximal parts of the peroneal veins and in the superficial femoral vein in the upper third of the thigh. The segment is unfilled in one film but fills out completely in a subsequent one (Fig. 5.3).

There are so many veins in the lower leg that it is quite common for some to remain unfilled, but the number can be reduced by injecting an adequate volume of contrast medium and by compressing the muscles distal to the underfilled segment to create a bolus of contrast medium (see Figs. 5.1 and 5.3).

Poor opacification of the deep veins of the calf can occur if the muscles are contracted by weight bearing on the foot during the examination (Fig. 5.4) as pointed out by Rabinov and Paulin.[9]

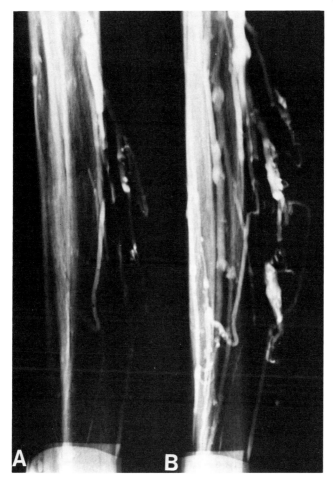

Fig. 5.4 (A) Underfilling of the calf muscle veins due to weight bearing during the examination. (B) There is better filling when the calf muscles are relaxed.

Mixing and layering defects

As mentioned above contrast medium is hyperbaric and tends to layer out resulting in uneven mixing,[4] (Fig. 5.5).

This phenomenon can give rise to a variety of artefactual appearances. A thin layer of contrast medium may hug the vein wall at the end of the injection so that the central stream of unopacified blood may closely resemble a recent thrombus in a single film (Fig. 5.6). Alternatively, the stream of contrast may be thicker on one side of the vein than on the other, again mimicking a thrombus undergoing retraction (Fig. 5.7). In its grossest form this lack of even opacification of a vein may closely resemble recanalisation changes from past venous thrombosis (Fig. 5.8).

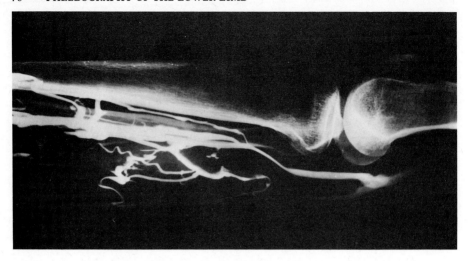

Fig. 5.5 A 'shoot through' lateral view of a calf phlebogram. Note the layering of the hyperbaric contrast medium in the calf veins, particularly in the popliteal vein.

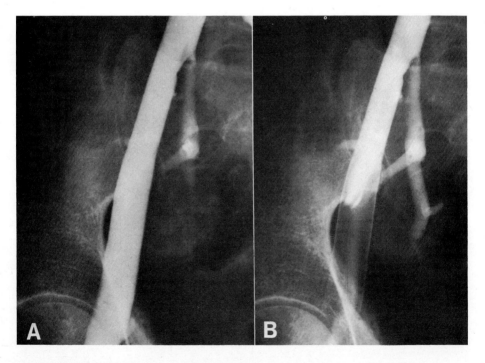

Fig. 5.6 (A) An early film in the series shows the vein is fully opacified. (B) A film taken at the end of the injection shows a thin layer of contrast medium around the edges of the blood stream which mimics the appearance of fresh thrombus in the external iliac vein. This appearance in a single film would be difficult to distinguish from thrombus so that at least two films of each part of the venous system should be taken in every examination.

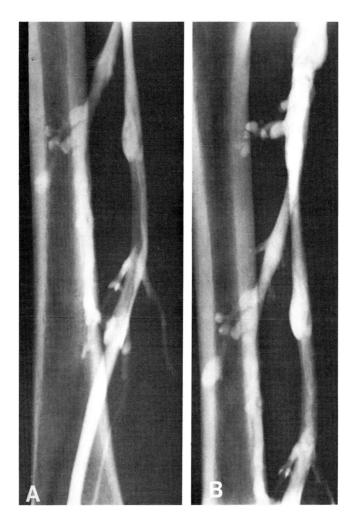

Fig. 5.7 (A) A long stream of poorly opacified blood in the superficial femoral vein which looks like a retracting thrombus. (B) This is not present in the next film.

These defects are usually eliminated by using a sharp foot-down table tilt to promote mixing, the use of sufficient contrast medium to fill the veins, the bolus technique to improve filling of the more proximal veins, and if necessary tourniquets to delay venous emptying supplemented if necessary by additional projections.[1,6]

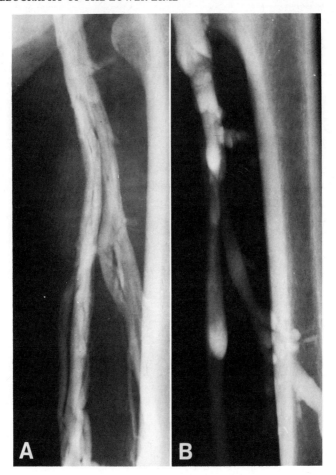

Fig. 5.8 (A) Gross failure of mixing resembles recanalisation changes in the superficial and deep femoral veins. (B) A later film during a Valsalva manoeuvre indicates that the veins are normal.

Entry of nonopacified blood

Nonopacified blood streaming into an opacified vein from tributaries can produce translucent defects: a common site for such a defect is where the large venous return of the profunda vein enters the opacified common femoral vein (Fig. 5.9). The appearances may be seen at any venous tributary and if the tributary is seen end on the appearance known descriptively as a 'knot hole effect' is produced (Fig. 5.10). The phenomenon can be detected because the appearances are not the same in every film and they are usually abolished by the use of the bolus technique or a Valsalva manoeuvre.

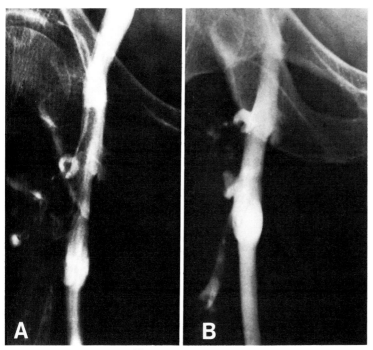

Fig. 5.9

Fig. 5.9 (A) A stream of non-opacified blood enters the common femoral vein from the profunda femoris vein giving appearances suggestive of a 'loose tail' of thrombus. (B) A film taken during a Valsalva manoeuvre fills out the vein.

Fig. 5.10 Small 'knot holes' are present in the lateral peroneal vein due to the entry of tributaries seen end on.

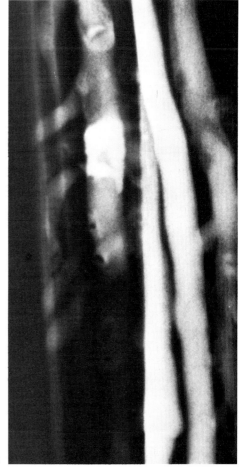

Fig. 5.10

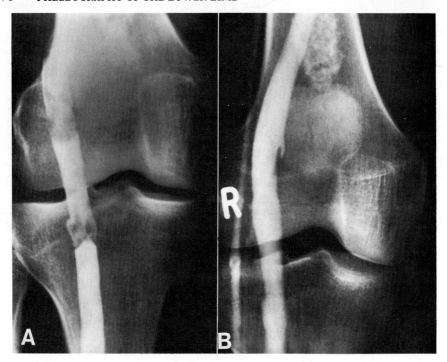

Fig. 5.11 (A) Poor mixing due to turbulence around two valves in the popliteal vein resembling thrombus. (B) The filling defects are abolished by better filling. There is an infarct in the lower femur.

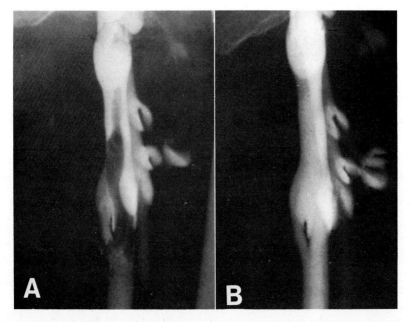

Fig. 5.12 (A) Streaming of contrast medium through a partially opened valve resembling a thrombus with a 'loose tail'. (B) A subsequent film shows the vein is normal.

Valve defects

Leaflets of the bicuspid venous valves can almost always be recognised and can be more clearly seen following a Valsalva manoeuvre.

Depending on the degree of valvular opening turbulent blood may create a filling defect in association with the valve; this defect however changes from film to film (Figs. 5.11 and 5.12).

Air bubbles

The inadvertent injection of air with the contrast medium produces easily recognisable filling defects which may vary in position between films and also with the position of the patient during examination (Fig. 5.13).

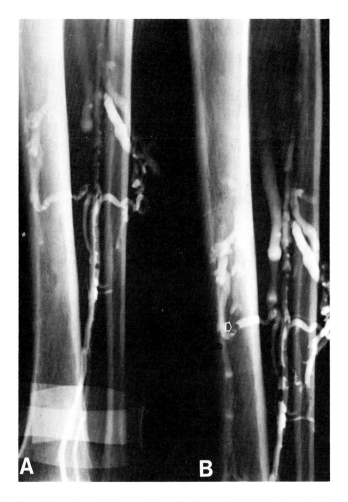

Fig. 5.13 (A) Air bubbles inadvertently injected. (B) The defects are spherical and have changed in position and shape in a later film. An arrow indicates one air bubble.

The Venturi effect

Under certain optimal physiological conditions the sides of a thin walled tube may be drawn together when a jet of fluid is injected into it. This phenomenon is sometimes seen in large veins when contrast is injected rapidly through an end opening cannula or needle. It causes a smooth tapering of the vein just distal to the cannula tip and may be mistaken for a smooth stenosis, but its appearance is transient (Fig. 5.14).

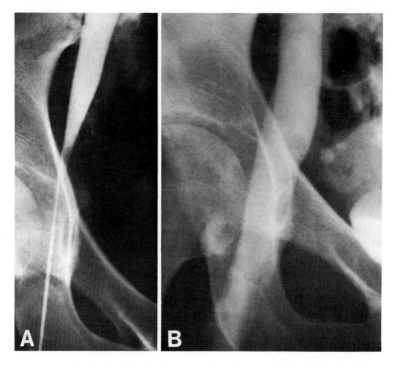

Fig. 5.14 (A) The vein just beyond the tip of the cannula is narrowed during the injection because of the Venturi effect 'sucking' the walls together. (B) A later film shows that this is an artefact.

The effect of posture

The degree of flexion or extension of the limbs may distort the calibre of the veins especially in relation to joints. Thus, hyperextension of the knee may narrow or totally occlude the popliteal vein,[2] (Fig. 5.15) and hyperextension of the hip increases the compression of the femoral vein by the inguinal ligament (Fig. 5.16). In the supine position the left common iliac vein may be compressed between the right common iliac artery in front and the lumbosacral spine behind, partially obstructing the vein. A change of posture may relieve the compression and the vein then appears normal (Fig. 5.17).

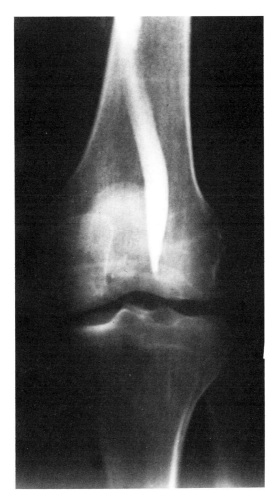

Fig. 5.15 Hyperextension of the knee during ascending phlebography completely occluding the popliteal vein. Other films with the knee in a neutral position showed that the popliteal vein was normal.

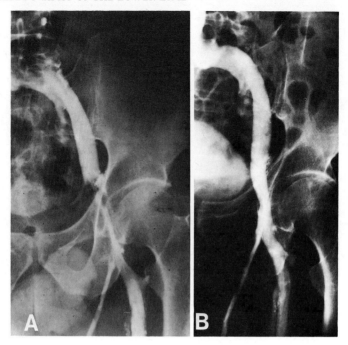

Fig. 5.16 (A) There is narrowing of the common femoral vein due to the inguinal ligament in hyperextension of the hip joint. In some patients this compression leads to thrombosis and occlusion. (B) In a relaxed position the vein is shown to be normal.

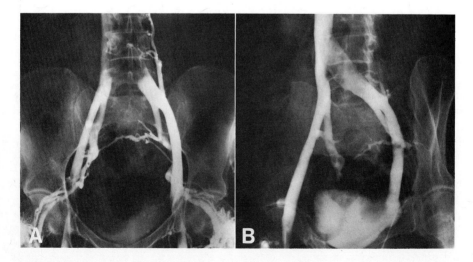

Fig. 5.17 (A) In the supine position the left common iliac vein appears to be obstructed. (B) A repeat examination with the patient turned slightly oblique shows no evidence of obstruction although the vein is flattened.

ARTEFACTS 79

Extrinsic pressure defects

Arterial impressions

The commonest site for arterial compression is at the termination of the left common iliac vein where it is crossed by the right common iliac artery (Fig. 5.18).

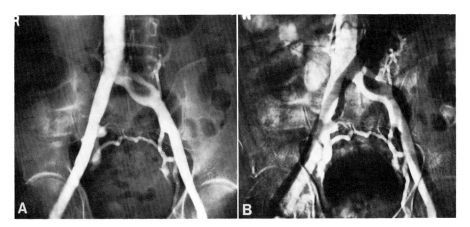

Fig. 5.18 (A) There is a rounded filling defect in the distal part of the left common iliac vein. (B) A superimposed subtraction film of an arteriogram shows that this is due to a dilated and tortuous left common iliac artery in a patient with arteriomegaly.

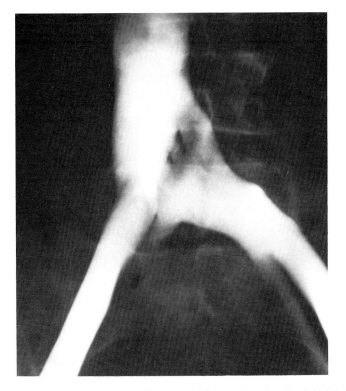

Fig. 5.19 Anteroposterior synechiae (webs) between the front and back walls of the left common iliac vein due to past thrombosis.

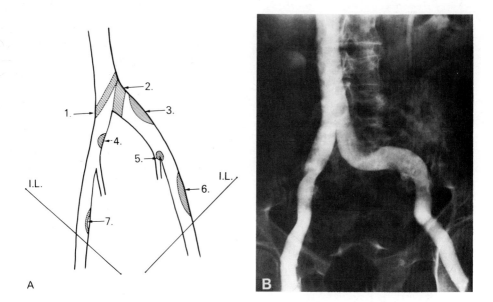

Fig. 5.20 (A) Diagram of the common sites where arteries may indent veins. (1) Right common iliac artery in an unusually high position. (2) Right common iliac artery in the usual position. (3) Left common iliac artery. (4) Right internal iliac artery. (5) Left internal iliac artery. (6) Left external iliac artery. (7) A branch of the right internal iliac artery. (B) Sites 3, 4, 5, 6 and 7 are compressed in this phlebogram.

Under certain circumstances, notably prolonged bed rest, stasis at this site may predispose to thrombosis and occlusion.[3,8] (Fig. 5.19).

There are many other sites where arterial impressions may occur,[7] (Fig. 5.20) particularly when the arterial tree is dilated as in the form of arteriosclerosis known as arteriomegaly.[5]

Other causes of external compression
Because the vein walls are thin and easily deformed by adjacent structures, such common conditions as spinal osteophytes, normal intervertebral discs, disc protrusions and other spinal deformities, abnormally positioned viscera and haematomas due to trauma or surgical interventions may distort the normal venous channels.

These lesions which give rise to external compression of the veins are discussed in Chapter 10.

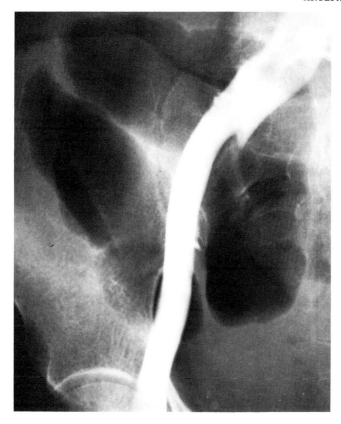

Fig. 5.21 There is a relative translucency in the common iliac vein due to bowel gas shadows. The translucent gas shadows extend beyond the opacified veins.

Overlapping shadows

Overlying shadows of bowel gas, faeces, viscera and bones, as well as overlap of opacified veins themselves all produce confusing appearances in some instances. Visceral and bone shadows or translucencies due to gas can usually be seen to extend beyond the confines of the veins (Fig. 5.21) but additional projections may be required if genuine confusion exists. Similar overlapping of contrast filled veins causing difficulty in interpretation may be prevented by rotation of the limb to separate the images of the bones and the veins (Fig. 5.22).

Iatrogenic artefacts

Tourniquets
The use of tourniquets may give rise to pressure artefacts in the contrast filled veins but provided the tourniquet is slightly opaque their site and any effect they may have on the vein can always be appreciated (Fig. 5.23). If there is any real doubt a film can be taken after the tourniquet has been released.

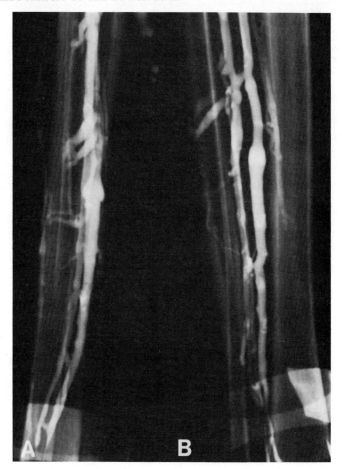

Fig. 5.22 (A) A lateral calf phlebogram. The peroneal veins overlap each other. (B) In the anterior projection with the foot internally rotated the veins are separated and shown to be normal. The patient had signs and symptoms suggesting deep venous thrombosis of the calf which the phlebogram excludes.

Plication
Plication of the inferior vena cava, or occasionally the common iliac or common femoral vein, is often employed to check thrombo-embolic disease. The operation distorts the outline of the vein and leaves a number of small defects at the site of the stitches (Fig. 5.24).

CONCLUSION

There is generally little difficulty in identifying the many artefacts which may arise in the venous system.

The most important differential diagnosis is that of venous thrombosis, the diagnostic criteria for which are discussed in Chapter 6. Most of the artefactual appearances are inconstant in a series of films and should not therefore be confused with thrombus.

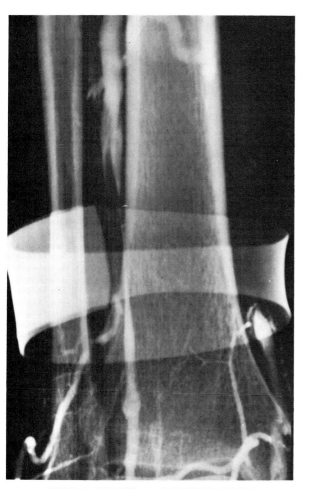

Fig. 5.23 The veins around the ankle are compressed by the tourniquet. The self fastening rubber tourniquet is slightly opaque showing its exact position.

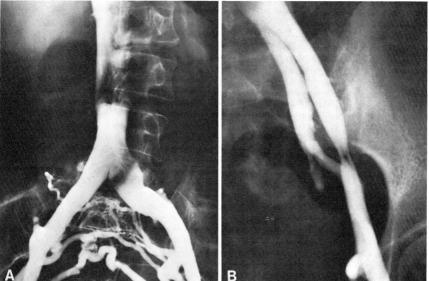

Fig. 5.24 (A) Plication of the inferior vena cava. (There is a marked iliac compression defect.) (B) Plication of the left external iliac vein.

The majority of difficulties arise from faulty technique. The main error is to use too little contrast medium delivered at too low a pressure causing mixing defects.

In the proximal parts of the venous system, or even more distally if valves are destroyed, the Valsalva manoeuvre is of value, not only for the identification of valves but also to slow down the flow of contrast opacified blood in the veins enabling them to be more clearly outlined. The bolus technique is also helpful to fill out veins with contrast. In addition steep table tilts promote mixing of the blood and contrast medium and the use of an ankle tourniquet to direct the contrast medium into the deep veins and an above knee tourniquet to delay emptying the calf veins also result in better opacification.

Lastly the fluoroscopic technique, recommended as the standard method for ascending phlebography in Chapter 3, enables films to be taken when the veins are seen to be fully opacified so that artefacts due to inadequate filling are less likely to occur than when a 'blind' overcouch technique is employed. Fluoroscopy also enables the radiologist to alter the position of the patient to produce clear, unobscured phlebograms.

REFERENCES

1. Almen T, Nylander L 1964 False signs of thrombosis in lower leg phlebography. Acta Radiologica 2: 345
2. Arkoff R S, Gilfillan R S, Burhenne H J 1968 A simple method for lower limb phlebography – pseudo-obstruction of the popliteal vein. Radiology 90: 66
3. Cockett F B, Lea Thomas M 1965 The iliac compression syndrome. Brit J Surg 53: 99
4. Kjellberg S R 1943 Die Mischungs- und Strömungsverhältnisse von wasserlöslichen Kontrastmitteln bei Gefäss- und Herzuntersuchungen. Acta Radiol 24: 433
5. Lea Thomas M, Andress M R 1970 Phlebographic changes in arteriomegaly. Acta Radiologica 10: 427
6. Lea Thomas M, Carty H 1975 The appearances of artefacts on lower limb phlebograms. Clin Radiol 26: 527
7. Lea Thomas M, Andress M R, Fletcher E W L 1968 Arterial impressions in pelvic phlebography. Clin Radiol 9: 404
8. Negus D, Fletcher E W L, Cockett F B, Lea Thomas M 1968 Compression and band formation of the mouth of the left common iliac vein. Brit J Surg 55: 24
9. Rabinov K, Paulin S 1972 Roentgen diagnosis of venous thrombosis of the leg. Arc Surg 104: 134

6

Thromboembolism

THROMBOGENESIS

The detailed mechanism of thrombogenesis and thrombolysis need not concern the phlebologist. Only an outline discussion of the mechanism is presented in this chapter; for a more detailed account readers are referred to the monograph by Hume and his colleagues.[13]

The initiating stimulus in venous thrombosis is usually at a site where local stasis has encouraged the concentration of activated clotting factors with the deposition of platelets over a small area. The thrombus is formed in flowing blood and has a layered structure containing varying amounts of red cells, groups of granular leucocytes and condensed masses of platelets bound by fibrin (Fig. 6.1).

The point of attachment of the thrombus may be very small but it may propagate for some distance without provoking an inflammatory reaction, and may become completely detached to embolise to the pulmonary circulation.

More frequently thrombus stimulates an inflammatory reaction in the vein wall so that it becomes adherent to the endothelium and in this way detachment and embolism is prevented.[13] Restoration of the lumen, often with valvular damage, is eventually achieved by thrombolysis or retraction and recanalisation.[5]

The most common sites of stasis are the valve cusp pockets (Fig. 6.2) the soleal sinusoidal veins, or where local damage has occurred to the vein wall following trauma, surgery or inflammation.

Although the majority of thrombi begin in the calf[19,26] some originate in the great veins while others arise from the foot veins and spread to the calf,[20] (Fig. 6.3). In many cases there is a multicentric origin.[4] Thrombus tends to propagate in the direction of the blood stream but it can propagate in both directions.

These pathological processes are reflected in the phlebographic appearances.

Phlebography in deep vein thrombosis

It has been estimated that phlebography is about 95 per cent accurate as a method of demonstrating peripheral venous thrombosis.[3]

When phlebography is undertaken to confirm the presence of deep vein thrombosis the standard technique of ascending phlebography should be followed with special emphasis on the following objectives:

(a) If a limb is found to contain thrombus the other limb must also be examined since there is a 50 per cent chance that it will also contain thrombus.

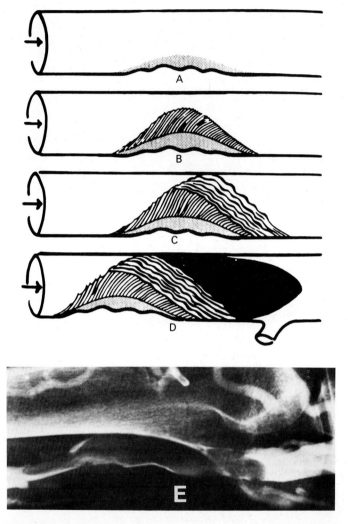

Fig. 6.1 The pathology of thrombus. (A) Platelet deposition. (B and C) The development of a laminated structure consisting of successive layers of fibrin and red cells. (D) The thrombus is propagating. (E) A phlebogram of a recent thrombus with a rounded proximal end as the thrombus propagates in the blood stream.

(b) The upper limit of any thrombus or occluded vein must be demonstrated if necessary with supplementary perfemoral or pertrochanteric injections.

(c) More than one film must be taken of each segment of vein to confirm that any filling defects are consistent and of constant shape.

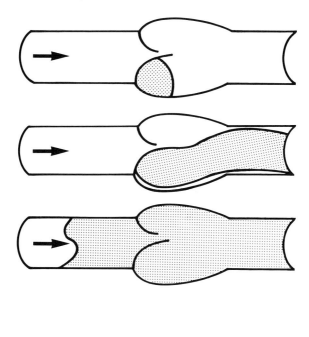

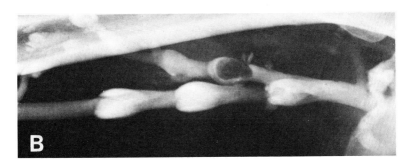

Fig. 6.2 (A) Diagram of propagation of thrombus from a valve cusp. (B) Recent thrombus in a valve cusp pocket in the profunda femoris vein.

(d) As many deep leg, thigh and pelvic veins as possible should be filled – especially any collateral channels which may bypass occluded veins.

(e) The ankle tourniquet must not be so tight that the superficial veins are completely obstructed, otherwise superficial vein thrombosis can be missed.

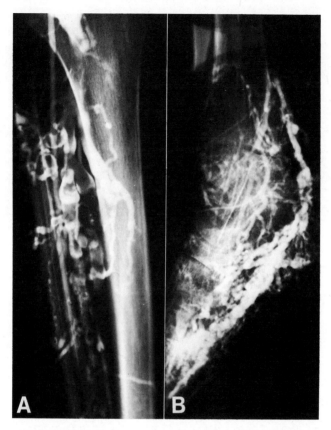

Fig. 6.3 (A) There is loose thrombus in the stem and muscle veins of the calf surrounded by a thin white line of contrast medium. (B) The veins of the foot are irregular with small thrombi in some of them. These changes are those of retraction and recanalisation and indicate that the thrombus in the foot is older than that in the calf. The thrombus may well have started in the foot veins and spread to the calf.

The appearances of venous thrombosis
Superficial thrombophlebitis
Superficial thrombophlebitis of the leg is usually secondary to chemical or bacterial inflammation of the vein wall. Phlebography plays little part in the diagnosis of this condition as the clinical signs are usually obvious. Superficial thrombosis is not usually of serious consequence to the patient and responds to local treatment. Its importance lies in the fact that it may propagate from the long or short saphenous veins into the deep veins (Fig. 6.4). It is for this reason that in the standard technique for ascending phlebography of the legs in suspected venous

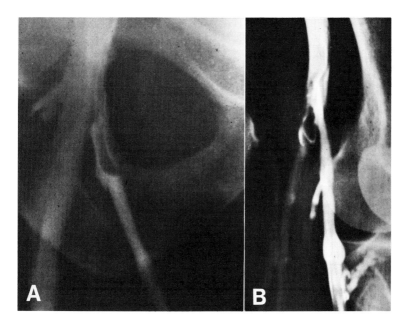

Fig. 6.4 (A) There is recent thrombus at the termination of the long saphenous vein. (B) This phlebogram shows recent thrombus in the termination of the short saphenous vein. Thrombi at these situations are not usually of significance although they may be the cause of minor pulmonary embolism. The main danger is that they may propagate into the deep veins resulting in thrombi of sufficient size to cause serious pulmonary embolism.

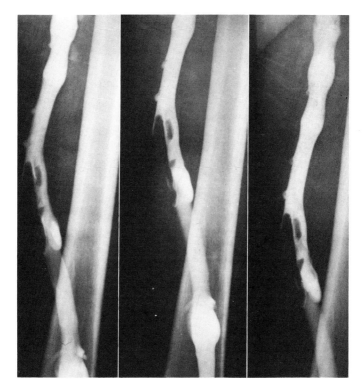

Fig. 6.5 Thrombus arising in the valve cusps of the superficial femoral vein. The appearance of the defects are identical in three radiographs taken with a time interval between them.

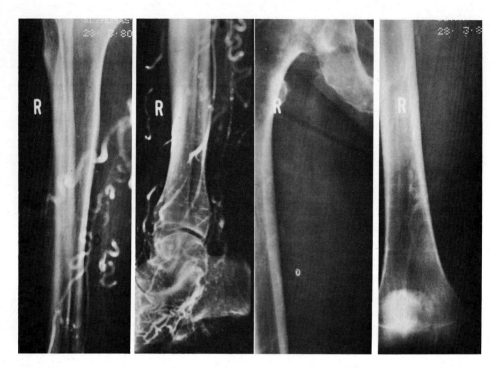

Fig. 6.6 Much of the deep venous system of the calf and leg is unfilled by contrast medium. This is not a technical fault as careful examination of the phlebogram shows a little thrombus in the posterior tibial vein (arrow) indicating that the deep venous system is almost totally occluded by thrombus.

thrombosis, complete occlusion of the superficial veins by the ankle tourniquet is not recommended.

The appearance of acute thrombosis in the superficial veins of the leg is similar to that of the deep veins.

Deep vein thrombosis
One of the main reasons why phlebography was slow in being generally accepted as a diagnostic technique was that poor quality phlebograms led to difficulty in interpretation, false positives and negatives being frequently reported. This poor diagnostic quality was largely the result of inadequate opacification of veins due to the use of insufficient contrast medium, and poor understanding of the possible artefactual appearances which may be confused with thrombus. Using the techniques described in Chapter 3 with a knowledge of the artefactual appearances

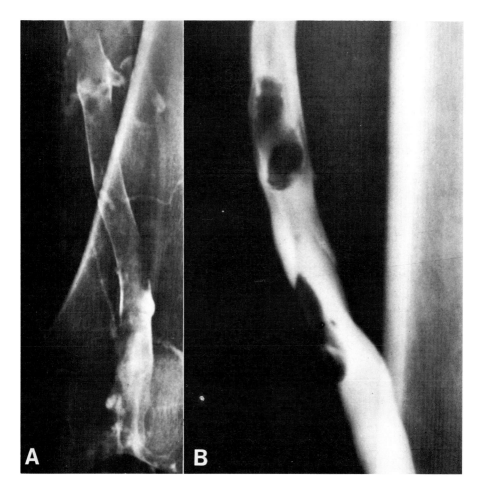

Fig. 6.7 (A) This recent thrombus is cylindrical in shape conforming to the shape of the femoral vein. (B) These older thrombi which are contracting and adhering to the femoral vein wall appear lobulated (upper thrombus) and plaque like (lower thrombus).

which occur and which are discussed in Chapter 5, it should be possible to diagnose the presence or absence of thrombus with a high degree of accuracy.

Venous thrombosis shows as a constant filling defect in an opacified vein and should have the same size and shape in at least two films with a short interval between them (Fig. 6.5). It is important for accurate diagnosis to try to outline the thrombus itself and not to rely solely on non-filling of veins. The veins should be carefully examined for evidence of thrombus within them (Fig. 6.6).

The shape of an intraluminal filling defect due to thrombus depends on the size

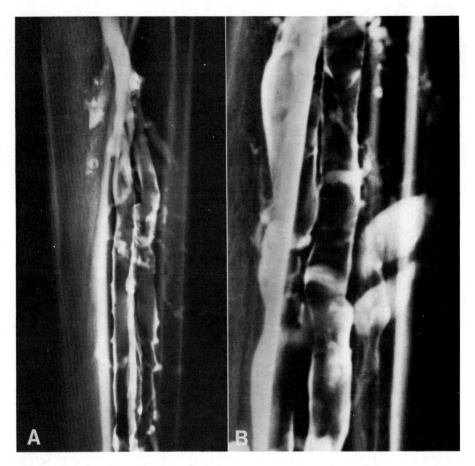

Fig. 6.8 (A) Recent thrombus in the stem veins of the calf. The thrombus is surrounded by a thin uniform line of contrast medium. (B) This is a close up view of part of the thrombus.

of the thrombus, its site of origin, its age and the projection in which it is radiographed (Fig. 6.7).

If a thrombus is very fresh it will not be adherent to the wall and will appear as a translucent defect separated from the wall by a thin white line of contrast medium and classically presents a 'ground glass' appearance (Figs. 6.8, 6.9, 6.10 and 6.11). Obliteration of this thin contrast line indicates adherence of the thrombus to the wall (Figs. 6.12, 6.13 and 6.14).

When a thrombus occludes a vein there is no contrast around it, but there is

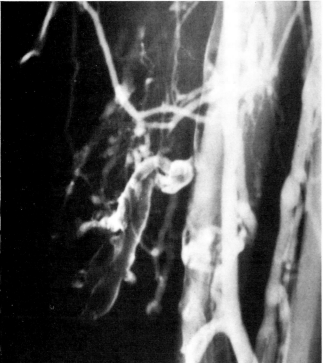

Fig. 6.9 A close up view of recent thrombus in a muscle vein of the calf. The thrombus is entirely surrounded by contrast medium.

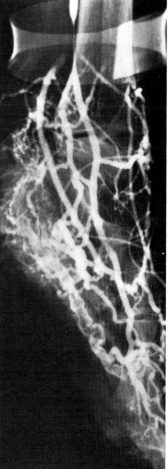

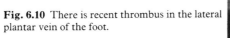

Fig. 6.10 There is recent thrombus in the lateral plantar vein of the foot.

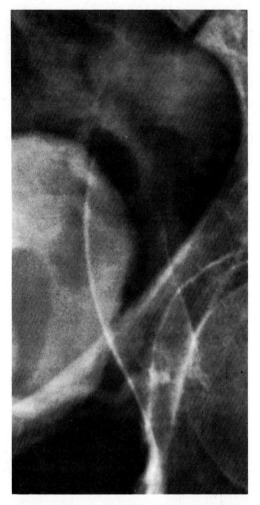

Fig. 6.11 This recent thrombus in the iliac vein shows the classical 'ground glass' appearance. It also shows a very thin clear line of contrast around the thrombus indicating that it is not adherent to the wall. The thrombus is less than three days old.

contrast medium above and below it and in the collaterals beside it. Unlike collateral arteries, collateral veins enlarge almost immediately the segment has become obstructed and their number and extent only give a rough indication of the duration of the obstruction. These collaterals may take the form of enlarged venae commitantes which may be recognised because they are always slightly

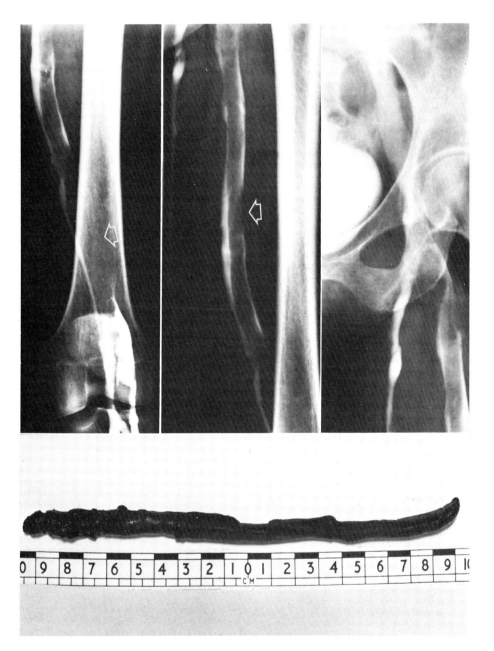

Fig. 6.12 The thrombus is recent but is beginning to adhere to the vein wall shown by obliteration of the surrounding contrast line in parts (arrows). There is also evidence of retraction indicated by a thicker line of contrast elsewhere. The thrombus is still too loose to be safely treated conservatively and was removed by thrombectomy. The operative specimen is shown in the photograph.

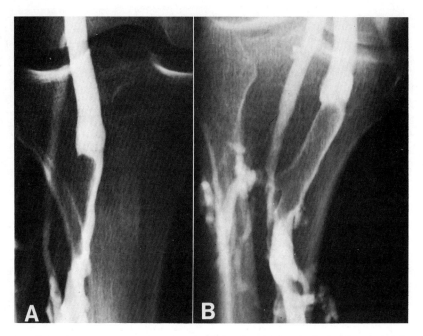

Fig. 6.13 This phlebogram shows the value of additional projections in doubtful cases. (A) A straight projection showing that the thrombus is adherent to the lateral wall. (B) In the oblique projection contrast surrounds the thrombus indicating that it is not completely adherent.

smaller but in the same line as the veins they replace (Fig. 6.15); other deep veins including muscular tributaries and also superficial veins may form a collateral pathway (Fig. 6.16).

As the thrombus ages it retracts and becomes smaller, allowing a thicker layer of contrast medium to surround it and its outline becomes more clearly defined (Fig. 6.17).

Finally the process of organisation and retraction of the thrombus, so called recanalisation, restores the lumen of the vein but this is left irregular, often reduplicated and with its valves damaged or destroyed (Figs. 6.18 and 6.19). The process of recanalisation does not always result in severely damaged veins; phlebographically normal appearances, or only minor changes may follow extensive thrombosis where thrombolysis is complete, whether this is the result of therapy or due to natural processes (Figs. 6.20, 6.21 and 6.22). It is not however possible to predict the outcome from the initial phlebogram.

In the first few hours after calf vein thrombosis, inflammatory reaction with

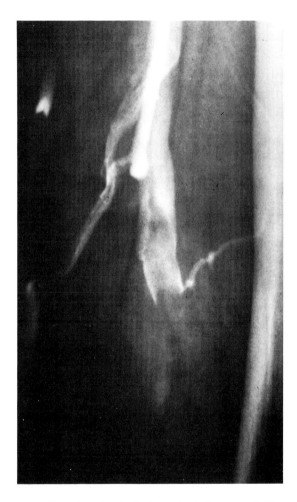

Fig. 6.14 The upper part of the thrombus in the femoral vein is surrounded by a thin white line indicating that it is loose. Distally the white line is totally obliterated indicating complete adherence.

oedema develops in the surrounding tissues causing a rise in pressure in the vascular compartment of the calf. This rise in perivenous pressure to levels above the venous pressure may, according to Nylander[27] prevent opacification of the deep veins. The author has not experienced this phenomenon, probably because an ankle tourniquet is always used to direct the contrast into the deep veins.

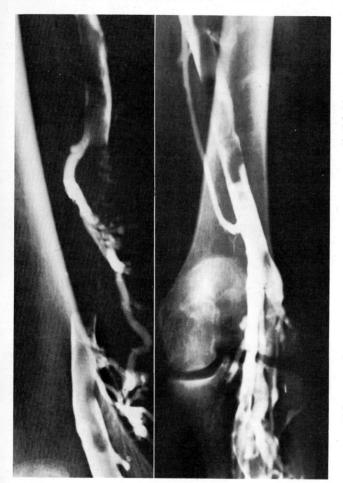

Fig. 6.15 Two examples of vena comitans functioning as collaterals bypassing recent femoral vein thrombotic occlusion.

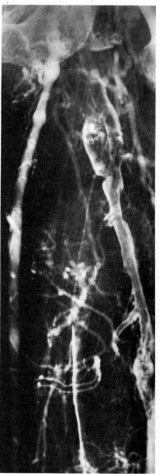

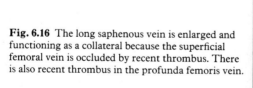

Fig. 6.16 The long saphenous vein is enlarged and functioning as a collateral because the superficial femoral vein is occluded by recent thrombus. There is also recent thrombus in the profunda femoris vein.

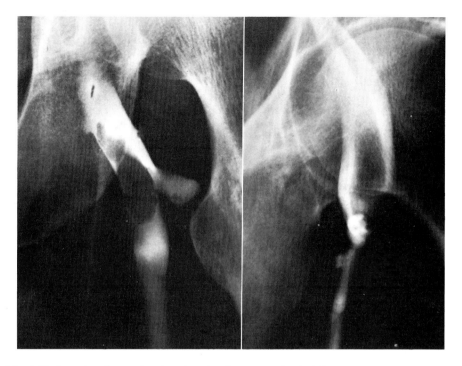

Fig. 6.17 Examples of retracting thrombus showing a more clearly defined outline and a thicker white line of contrast surrounding them. The 'ground glass' appearance is no longer present. The thrombi are approximately 10 to 14 days old.

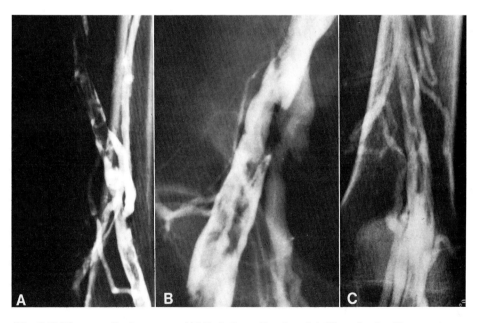

Fig. 6.18 The recanalisation process. (A) Early Stage (2 to 3 weeks). Thrombus is still present but the lumens of the popliteal and femoral veins are irregular and no valves can be identified. The femoral vein is totally occluded in the adductor canal. (B) Later stage (4 to 6 weeks). In this example some organising thrombus remains but the lumen is very irregular. (C) Final stage (3 months). The popliteal and femoral veins contain no thrombus but are irregular, reduplicated, and have destroyed valves. Collateral veins are present.

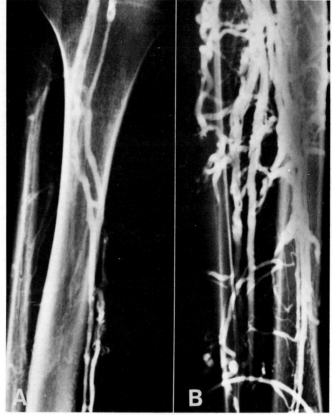

Fig. 6.19

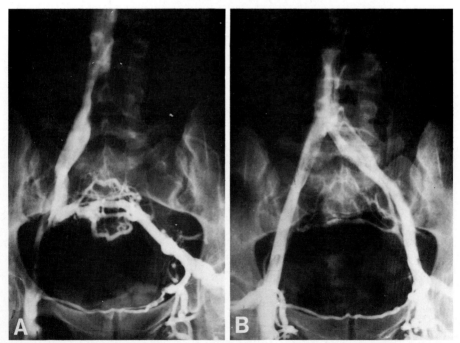

Fig. 6.20

Estimation of the age of thrombus

Before the fibrinogen uptake test became readily available it was shown that the phlebographic appearances of thrombus could be correlated with the duration of symptoms and signs in the leg.[18] Such a correlation is necessarily crude because many thrombi remain symptomless for a variable time and the clinical diagnosis is notoriously inaccurate. Nevertheless the study suggested that in the first week the thrombus is smooth in outline, virtually fills the whole of the vein allowing a thin white line of contrast medium to surround it. The contrast line may be obliterated in parts indicating adherence to the wall at these sites.

Over the next 10 to 14 days the thrombus becomes more adherent to one side of the vein wall and then begins to retract resulting in a clearer margin to the thrombus and a thicker layer of contrast medium around it. Further retraction and resorption results in recanalisation changes which progress over several months. If the recanalisation process does not occur, the vein remains permanently occluded and bypassed by collaterals (Fig. 6.23).

The detailed phlebographic appearances of thrombus have already been described earlier in this chapter and the results of thrombosis of the veins are discussed in Chapter 7.

PULMONARY EMBOLISM

Incidence

The overall incidence of deep vein thrombosis based on post mortem studies of specified groups of medical and surgical patients is estimated to be between 34 per cent and 50 per cent respectively.[11, 14, 24] For elderly patients and in the presence of trauma, figures of 80 per cent to 86 per cent have been quoted.[29] Using the most accurate method of detection of venous thrombosis – the fibrinogen uptake test, in the post-operative period, an incidence ranging from 50 per cent in patients with gynaecological operations[2] to 75 per cent in patients with pertrochanteric fractures[90] has been shown. In general surgical patients the incidence has been variously estimated from 24 per cent[26] to 35 per cent.[10] It must be assumed however that many small thrombi are of no clinical significance, being absorbed

---

Fig. 6.19 (A) The peroneal veins are occluded by recent thrombus and there is some recent thrombus in the popliteal vein. (B) Repeat phlebogram one year later shows that the peroneal veins are now patent although the lumens are irregular, the valves are destroyed and there are incompetent communicating veins. This is the post thrombotic state.

Fig. 6.20 (A) Initial phlebogram in a patient with phlegmasia cerulea dolens. There is total occlusion of the left external and common iliac veins and adherent thrombus in the inferior vena cava. The venous return from the left leg is through an enlarged internal iliac vein. Thrombectomy proved impossible. (B) Repeat phlebogram one year later. The iliac veins and the lower inferior vena cava are mildly irregular and there are a few pubic and pre-sacral collateral veins. The patient was symptom free ten years after the initial episode.

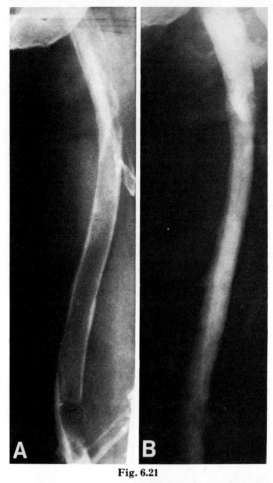

Fig. 6.21

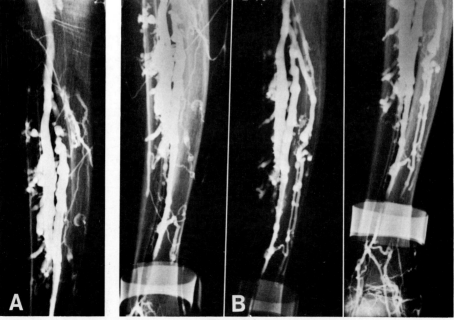

Fig. 6.22

by the natural processes of the body without the complications of embolism or post thrombotic sequelae.

An assessment of the mortality due to pulmonary embolism is difficult to obtain with accuracy but it is of the order of 3.7 per 10 000 patients.[23] The Registrar-General's figures of deaths attributable to pulmonary embolism are probably a gross underestimate not only because the diagnosis is often not verified by post mortem studies but also because of the method of classification. The Registrar-General's Statistical Review for England and Wales between 1941 and 1967[28] has been analysed in more detail by Hume et al[13] who considered that the true incidence of fatal embolism is in the region of about 21 000 per year, whereas the official figure is only 4981.

Accurate figures are also difficult to obtain in the United States of America, although it has been estimated that thromboembolism is responsible for the hospitalisation of 300 000 patients annually of which more than 5000 die.[15]

Despite the advances in diagnosis and management of pulmonary embolism the number of deaths recorded continues to increase (Fig. 6.24). Many patients who die from pulmonary embolism are elderly or suffering from terminal disease but it has been estimated that up to 60 per cent of patients having a fatal embolus would have carried on a normal life had it not been for this complication.[8, 25]

It is clear that the problem of pulmonary embolism is a major one.

Phlebography

The search for the source of the pulmonary embolus should include an ascending phlebogram of *both* limbs using the standard technique. Views of the feet should be taken and every attempt made to fill the profunda femoris and internal iliac veins and lower inferior vena cava (Fig. 6.25). If it is thought essential to show the whole of the internal iliac systems a bilateral pertrochanteric intraosseous phlebogram, in the supine position so that these veins are dependent, may be required.[16] Additional methods that may be required in some patients are retrograde phlebography[7] and cross over iliac phlebography.[6] For absolute completeness an angiocardiogram to show the right atrium, ventricle and pulmonary arteries may be necessary.

Barker and Priestly[1] reviewing the natural history of 1655 cases of post-operative thromboembolism found that 30 per cent of patients who survived one pulmonary embolism had a second embolus which proved fatal in 19 per cent of patients.

Fig. 6.21 (A) Initial phlebogram showing recent thrombus in the femoral vein.
(B) After only one week of anticoagulant therapy the vein is now phlebographically normal with valves demonstrated.

Fig. 6.22 (A) The posterior tibial veins are occluded by thrombus. (B) After a standard three day course of streptokinase the veins appear normal. It is hoped that such therapy will preserve valve function.

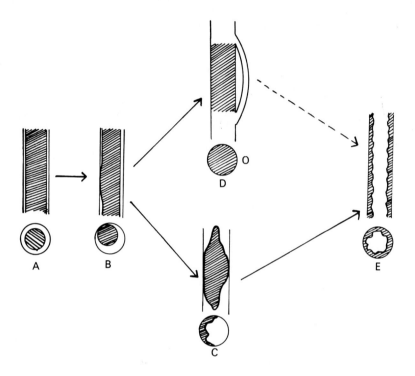

Fig. 6.23 A diagram of the progression of venous thrombosis. (A) Loose. (B) Adherent. (C) Adherence and retraction. (D) Occlusion with collaterals. (E) Recanalisation.

It is in the prevention of recurrent pulmonary embolism that phlebography plays such an important part by demonstrating if thrombus remains and whether it is likely to embolise.

Browse et al[5] found a statistically significant reduction in the recurrence rate of thromboembolism when treatment was based on phlebographic findings. In this series of 50 control patients treated by anticoagulation and without prior phlebography 13 (26 per cent) had a second embolism while on anticoagulants and 7 died, giving a death rate of 14 per cent. Another 7 per cent had emboli after stopping anticoagulants giving a total recurrence rate of 47 per cent.

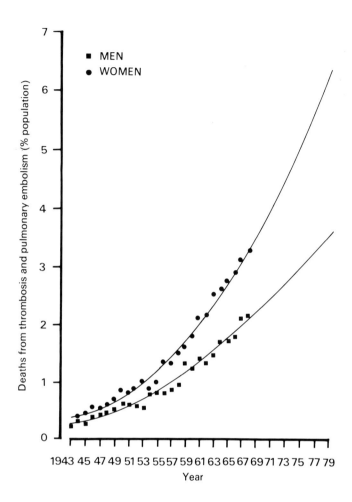

Fig. 6.24 Deaths from pulmonary embolism based on the Registrar General's Report for England and Wales 1943 to 1969, each curve is extrapolated to 1979 (After Negus, 1974)[25].

An important aspect of phlebography in patients with suspected pulmonary embolism is to assess whether the thrombus is loose or fixed firmly to the wall; additional projections such as oblique or lateral views may thus be required in doubtful instances. Occasionally a thrombus may appear loose in one projection but adherent in another (see Fig. 6.13). In the upper parts of the thigh and pelvis lateral projections are difficult to interpret as overlying bone and soft tissues tend to obscure contrast filled veins. Similarly with simultaneous bilateral injections there is overlap of the veins on each side. Oblique projections overcome these difficulties.

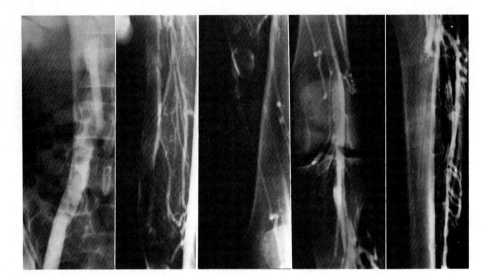

Fig. 6.25 A phlebogram in a patient with suspected pulmonary embolism. The whole of the deep venous system of the lower limbs has been examined. There is thrombus from the left calf to the lower inferior vena cava. The right leg (not illustrated) was free of thrombus.

Fortunately retraction and adherence tend to go together and it can usually be assumed that a retracted thrombus is also adherent to the vein wall and will not embolise. It can also usually be safely assumed that if no contrast medium passes from the leg veins into the large proximal veins, the thrombus is adherent to the vein wall and unlikely to embolise.

An important feature of the thrombus is its proximal extremity, the appearance of which provides information regarding the likelihood of embolisation. It may have a 'floating tail' (Fig. 6.26) which may become detached and embolise, or it may have a horizontal 'square cut' shape indicating that a portion has already broken off and embolised (Fig. 6.27).

While it is agreed that most fatal pulmonary emboli arise from the ilio-femoral segment or above[21,22] the size of the thrombus at any site must be taken into account. Occasionally the calf veins may be unusually capacious and if filled with

THROMBOEMBOLISM 107

Fig. 6.26 An example of a 'floating tail' of thrombus likely to embolise.

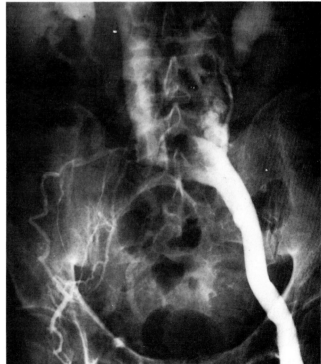

Fig. 6.27 An example of the 'square cut' appearance of the proximal end of the thrombus in the right common iliac vein indicating a piece has broken off and embolised.

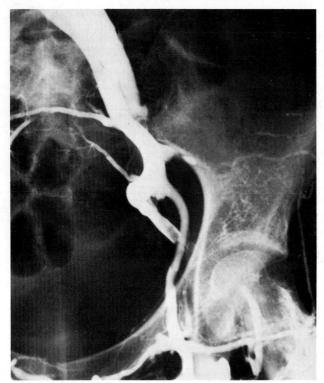

Fig. 6.28 Intraosseous pertrochanteric phlebogram in the supine position to show the internal iliac veins. This thrombus in the left internal iliac vein would not produce a fatal pulmonary embolus but repeated small emboli can produce pulmonary hypertension and death.

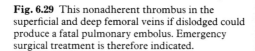

Fig. 6.29 This nonadherent thrombus in the superficial and deep femoral veins if dislodged could produce a fatal pulmonary embolus. Emergency surgical treatment is therefore indicated.

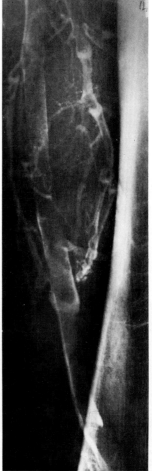

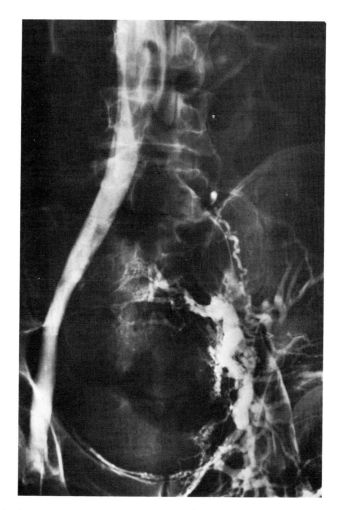

Fig. 6.30 A fresh, partly adherent thrombus is present in the upper part of the left femoral vein, the external iliac vein and the common iliac vein. This is unlikely to embolise. There is a large 'coiled tail' of thrombus lying free in the inferior vena cava. This patient had an inferior vena caval plication to contain this potential embolus. This phlebogram emphasises the importance of always showing the upper extent of a thrombus. The examination is a combined percutaneous femoral phlebogram and an intraosseous pertrochanteric phlebogram.

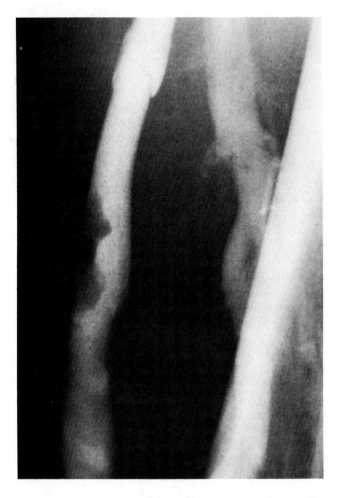

Fig. 6.31 This thrombus in the femoral vein is adherent and can safely be treated conservatively.

recent non-adherent thrombus, pulmonary embolism could prove fatal. Even small emboli can have serious consequences depending on the patients lung function which may be compromised from previous emboli or from co-existing respiratory disease (Fig. 6.28).

In summary, phlebography not only confirms the presence of a thrombus but its size and exact location, whether it is lying free within the lumen and therefore

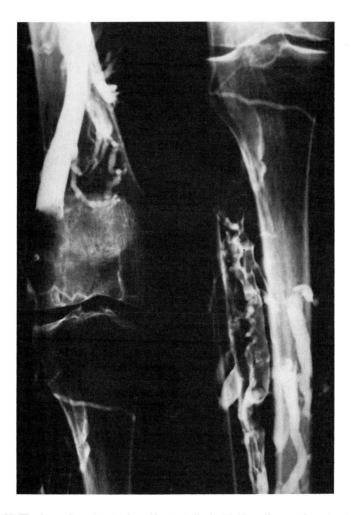

Fig. 6.32 The loose thrombus in the calf veins is 'locked in' by adherent thrombus in the popliteal vein. This thrombus cannot therefore embolise. Note the 'square end' of the popliteal thrombus indicating that an embolus has already occurred from this site.

liable to embolise or whether it is attached to the wall and unlikely to embolise (Figs. 6.29, 6.30 and 6.31). Phlebography will also indicate if loose distal thrombus is being 'contained' by more proximal adherent thrombus thus preventing embolism. In this situation the patient has performed his own auto ligation (Fig. 6.32).

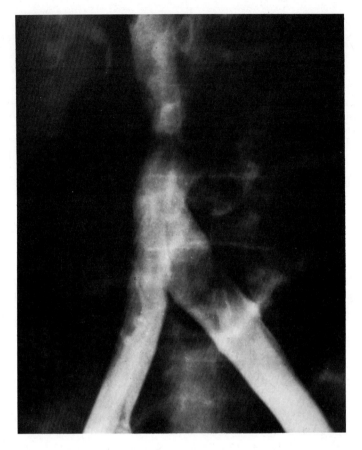

Fig. 6.33 Phlebogram showing inferior vena caval plication to prevent embolism from peripheral venous thrombosis.

Management
A large loose thrombus may need to be contained by plication of a more proximal vein (Fig. 6.33) or removed by thrombectomy, whereas an adherent thrombus may be treated conservatively with anticoagulation in the knowledge that there is no immediate risk of embolism (Fig. 6.34). Thrombolytic therapy may be used when the thrombus lies in free flowing blood. In this way it may be possible to lyse adherent thrombus and thus prevent or minimise damage to the deep venous system which otherwise might result in the post thrombotic complications discussed in Chapter 7.

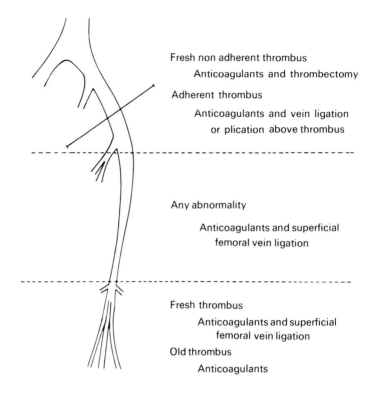

Fig. 6.34 Diagram to indicate the way in which phlebography may be used in determining the management of deep venous thrombosis (From Browse et al, 1969)[5].

PHLEGMASIA CERULEA DOLENS

Grégoire[12] first described the syndrome of massive venous occlusion of the limb producing oedema, intense cyanosis and temporary absence of arterial pulses. This syndrome is now commonly referred to as phlegmasia cerulea dolens to distinguish it from the commoner but less severe phlegmasia alba dolens. The foot and lower leg become cold, blue and swollen, there is considerable pain, and in the most severe cases so called venous gangrene develops.

The phlebogram shows massive iliofemoral thrombosis (see Fig. 6.20a). The most widely accepted explanation for the absence of arterial pulses in this condition is the greatly increased extra-arterial pressure due to massive venous outlet obstruction. In one case report[17] the arteries in the calf were small and occluded distally at the time of the onset of the phlegmasia (Fig. 6.35) but returned to normal when the venous thrombosis had subsided.

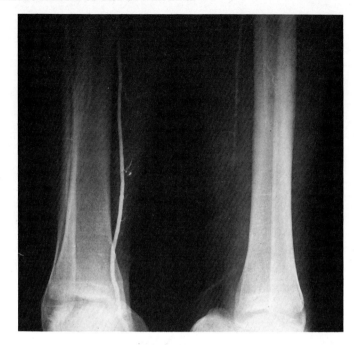

Fig. 6.35 Arteriogram of the calves in a patient with left phlegmasia cerulea dolens. The anterior tibial and the peroneal arteries in the left calf do not fill and the posterior tibial artery is small and becomes occluded in the lower calf. The calf arteries became normal on a repeat arteriogram two weeks later.

REFERENCES

1. Barker N W, Priestly J T 1942 Postoperative thrombophlebitis and embolism. Surgery 12: 41
2. Bonnar J, Walsh J 1972 Prevention of thrombosis after pelvic surgery by Dextran. Lancet 70, I: 614
3. Browse N L 1978 Diagnosis of deep vein thrombosis. Brit Med Bull 34: 163
4. Browse N L, Lea Thomas M 1974 Source of non-lethal pulmonary emboli. Lancet 1: 259
5. Browse N L, Lea Thomas M, Solan M J, Young A E 1969 Prevention of recurrent pulmonary embolism. Brit Med Bull 5, 3: 382
6. Buist T A S 1975 In: Ruckley C V and MacIntyre I M C (eds) Venous thromboembolic disease. Churchill Livingstone, Edinburgh
7. Dow J D 1973 Retrograde phlebography in major pulmonary embolism. Lancet 2: 40
8. Evans D S 1971 Early diagnosis of thromboembolism by ultrasound. Am Roy Coll Surg 49: 225
9. Field E S, Nicolaides A N, Kakkar V V, Crellin R Q 1972 Deep vein thrombosis in patients with fractures of the femoral neck. Brit J Surg 59: 377
10. Flanc C, Kakkar V V, Clarke M B 1968 The detection of venous thrombosis of the legs using ^{125}I-labelled fibrinogen. Brit J Surg 55: 742
11. Gibbs N M 1967 Venous thrombosis of the lower limbs with particular reference to bed rest. Brit J Surg 45: 209
12. Gregoire R 1938 Le phlebite bleue (Phlegmasia cerulea dolens) Presse med 2: 1313
13. Hume M, Sevitt S, Thomas D P 1970 Venous thrombosis and pulmonary embolism. Oxford University Press
14. Hunter W C, Kryger J J, Kennedy J C, Sneedon V D 1945 Etiology and prevention of thrombosis of the deep leg veins. Surgery 17: 178
15. Kakkar V V 1977 Diagnosis of deep vein thrombosis and pulmonary embolism. Triangle 16: 1
16. Lea Thomas M, Browse N L 1972 Internal iliac vein thrombosis. Acta Radiol Diag 12: 660
17. Lea Thomas M, Carty H 1975 Arteriographic changes in phlegmasia cerulea dolens. Aust Radiol 19: 57

18. Lea Thomas M, McAllister V 1971 The Radiological progression of deep vein thrombosis. Radiology 99: 37
19. Lea Thomas M, O'Dwyer A 1977 Site of origin of deep vein thrombosis in the calf. Acta Radiol Diag 18: 418
20. Lea Thomas M, O'Dwyer J A 1978 A phlebographic study of the incidence and significance of venous thrombosis of the foot. Radiology 99: 37
21. Mahaffy R G, Mavor G, Galloway J M D 1971 Iliofemoral phlebography in pulmonary embolism. Brit J Radiol 44: 172
22. Mavor G E, Galloway J M D 1967 The iliofemoral venous segment as a source of pulmonary emboli. Lancet 1: 34
23. Mortality statistics. England and Wales 1976 Her Majesty's Stationery Office, London
24. McLachlin J, Paterson J C 1951 Some basic observations on venous thrombosis and pulmonary embolism. Surg Gynecol 93: 1
25. Negus D 1974 The diagnosis and management of venous thrombosis. Teach-In 3: 9
26. Nicolaides A N, DuPont P A, Desai S, Lewis J D, Douglas J N, Fourides G, Luck R J, Jamieson C W 1972 Small doses of subcutaneous heparin in preventing DVT after major surgery. Lancet 2: 890
27. Nylander G 1968 Phlebographic diagnosis of acute deep leg thrombosis. Acta Chir Scand 397 (suppl): 30
28. Registrar-General's Statistical Review for England and Wales 1941–1967. Her Majesty's Stationery Office, Part I Medical, London
29. Sevitt S, Gallagher N G 1961 Venous thrombosis and pulmonary embolism: A clinico-pathological study in injured and burned patients. Brit J Surg 48: 475

7

Post thrombotic states

INTRODUCTION

Thrombosis of the deep veins of the leg is often followed after an interval of months or years by induration, pigmentation, ulceration, swelling and pain in the affected limb.

The most troublesome component of this syndrome is the venous ulcer of the ankle. For many years these were known as varicose ulcers and were considered to be always associated with primary varicose veins due to incompetent superficial veins. This idea was challenged as long ago as 1868 by Gay[11] who introduced the term post thrombotic syndrome and divided the ulcers of the leg into those associated with varicose veins treated by their removal, and those intractable ulcers which result from damage of the deep and communicating veins which may follow thrombosis.

The post thrombotic syndrome is not usually associated with gross superficial varicose veins although varicose veins of the secondary type due to destruction of the valves in the communicating veins do occur (see Ch. 8). It is difficult to be certain how many patients with deep venous thrombosis go on to develop post thrombotic changes. In some patients the thrombus will be lysed by the normal processes of the body and many patients with post thrombotic sequelae give no previous history of deep venous thrombosis. This is not surprising as it is known that as many as one third of deep venous thromboses do not develop symptoms or signs. It has been suggested that within 5 years of a deep vein thrombosis of the leg about half of the patients develop post thrombotic changes at the ankle.[7]

It is to be hoped that the treatment of deep vein thrombosis with anticoagulants to prevent propagation of thrombus from the calf into the more proximal veins and the use of thrombolytic agents will reduce the incidence of post thrombotic sequelae.

The morbidity caused by the post thrombotic syndrome is considerable. It has been suggested that as many as 0.5 per cent of the population of Great Britain and the United States of America may be affected.[4] In an epidemiological survey of 3000 gainfully employed persons, Widmer[31] found a morbidity related to symptoms of past thrombosis in 37.4 per cent women and 31.3 per cent men. The prevention and management of the syndrome is of considerable social and economic significance and phlebography plays an important part.

Following thrombosis of the deep veins of the calf, recanalisation almost

POST THROMBOTIC STATES 117

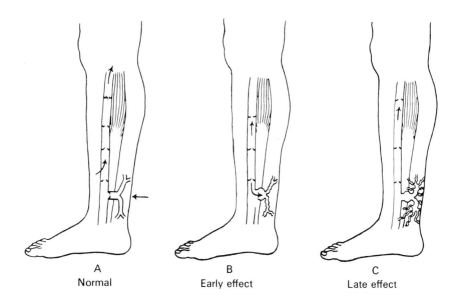

Fig. 7.1 Development of a venous ulcer. The pressure from each calf contraction is transmitted through incompetent ankle communicating veins to the superficial veins.

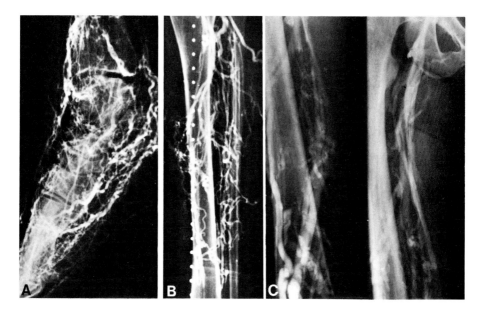

Fig. 7.2 Post thrombotic changes. The veins are irregular, reduplicated, collaterals are present and no normal valves can be seen. Incompetent communicating veins are present in the calf. (Note the ruler with 1cm metal markers to relate the sites of the incompetent communicating veins to the patient's leg. (A) Foot (B) Calf (C) Thigh

118 PHLEBOGRAPHY OF THE LOWER LIMB

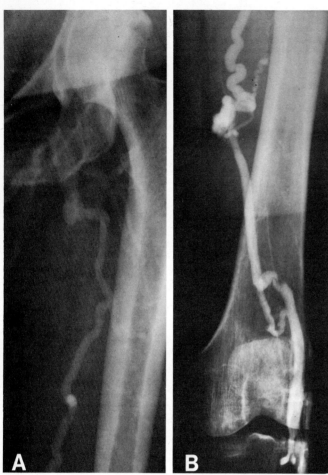

Fig. 7.3

Fig. 7.3 (A & B) A vena commitans functioning as a collateral. The superficial femoral vein is totally occluded. The vena commitans is tortuous and smaller than the segment of vein it replaces but lies in the same anatomical line.

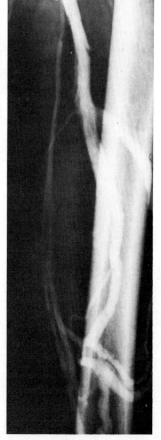

Fig. 7.4

Fig. 7.4 The superficial femoral vein is totally occluded and replaced by a small vena commitans. The profunda femoris vein is enlarged and functioning as a collateral.

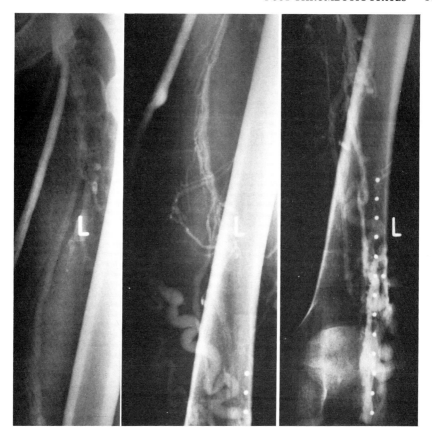

Fig. 7.5 The long saphenous vein is functioning as a collateral in occlusion of the superficial femoral vein.

invariably occurs,[2,17] but in the process the delicate valves are damaged and become incompetent.[8] Permanent stenosis or occlusion of a major vein may result, but an adequate venous return is usually re-established, at any rate for some years, through dilated collaterals.[12]

Either situation distorts the normal function of the calf muscle pump and the high pressure developed during exercise in the posterior tibial, peroneal and muscular veins is transmitted to the superficial veins through incompetent communicating veins (Fig. 7.1). This high pressure, probably combined with additional factors such as tissue response eventually causes skin necrosis and ulceration, usually on the medial side of the ankle but it may also occur on the lateral aspect or circumferentially.

Phlebographic techniques

Phlebography is particularly valuable in deciding the correct treatment. For example, in the presence of gross post-thrombotic changes surgery and sclerotherapy to the superficial veins is of doubtful value.

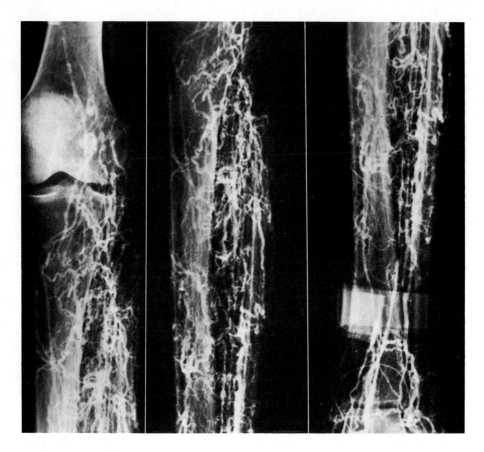

Fig. 7.6 Extensive post thrombotic changes in the calf. There are occlusions of parts of the stem veins of the calf and also of the popliteal vein. A very large number of collateral veins are present, arising from the deep and superficial venous systems and also the communicating veins.

The methods of phlebography used in investigating the post thrombotic states have been discussed in detail in Chapter 3. They include the standard method of ascending phlebography, descending phlebography and iliocaval phlebography, either by percutaneous or intraosseous techniques. Occasionally, retrograde phlebography using a brachial approach is required to show the upper end of an inferior vena caval occlusion.

The modifications of ascending phlebography to demonstrate incompetent communicating veins are described in Chapter 8. Ascending functional ciné phlebography[9, 14] allows the detailed function of the valves of the deep veins to be assessed more accurately. Ciné film, however, is an inconvenient form of radiography and much of the information provided can be obtained by careful fluoroscopy and spot films.

POST THROMBOTIC STATES 121

Fig. 7.7 There is a large incompetent communicating vein connecting with a varicose vein. The deep veins appear phlebographically normal.

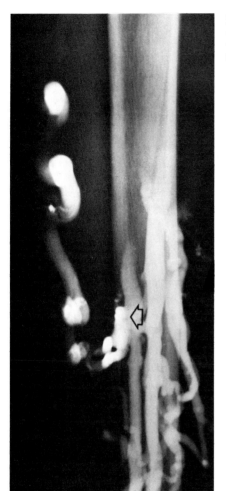

Fig. 7.7

Fig. 7.8 The deep venous system of the calf is irregular with destroyed valves indicating post thrombotic changes. There is a large incompetent communicating vein on the medial side of the lower calf connecting with a varicose vein.

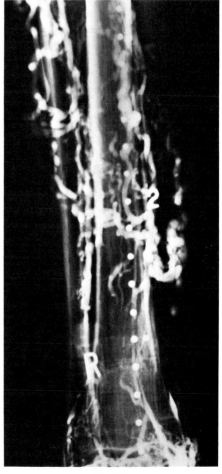

Fig. 7.8

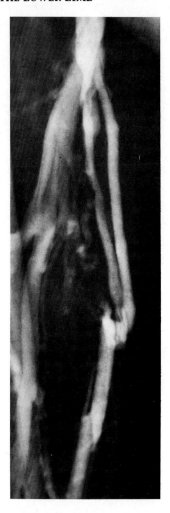

Fig. 7.9 Post thrombotic occlusion of the superficial femoral vein. The vein above and below the occlusion appears normal but there are some post thrombotic changes in the profunda femoris vein.

PHLEBOGRAPHIC APPEARANCES IN POST THROMBOTIC STATES

When a vein reopens as a result of retraction of the thrombus, so called recanalisation, the lumen is irregular, often reduplicated and the valves are damaged or destroyed (Fig. 7.2).

Collateral pathways develop which may originate in several ways. When venae commitantes form collaterals they may be recognised because they are slightly smaller but in the same line as the veins they replace (Fig. 7.3). Further pathways which may develop include other deep veins (Fig. 7.4), superficial veins (Fig. 7.5) or a combination of both.

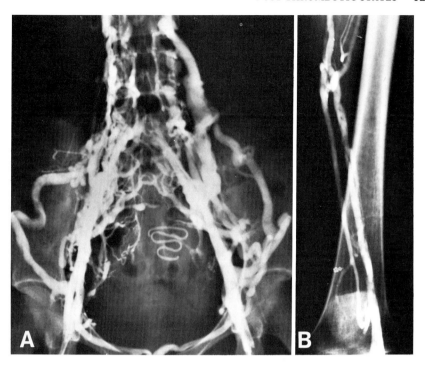

Fig. 7.10 Combination of destruction and valve incompetence. (A) The inferior vena cava is occluded with collateral formation. (B) The superficial femoral vein is narrowed, irregular, bypassed by collaterals and the valves are destroyed.

Occasionally incompetent communicating veins become part of the collateral circulation (Fig. 7.6).

There are four types of the post thrombotic state:

1. Incompetent communicating veins with normal deep veins (Fig. 7.7).
2. Incompetent communicating veins with damaged and incompetent valves in the deep veins (Fig. 7.8).
3. Various degrees of deep vein obstruction (Fig. 7.9).
4. Combinations of obstruction and valve incompetence (Fig. 7.10).

These post thrombotic sequelae are discussed below according to the different sites of occurence.

Peripheral leg veins

It is generally thought that thrombotic changes affect the valves of the stem veins of the calf as well as the communicating veins, however phlebographically incompetent communicating veins are often demonstrated with deep veins which appear normal. When deep veins are severely damaged they show typical changes

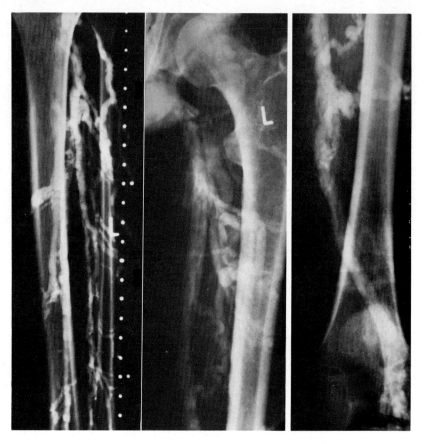

Fig. 7.11 Severe recanalisation changes affecting the leg veins. No valves can be identified and they are presumed to have been destroyed by the thrombotic process.

of recanalisation and the valves can be seen to be damaged or are not visible and are presumed to be totally destroyed (Fig. 7.11). Valve function is best shown phlebographically by descending phlebography but ascending phlebography with a Valsalva manoeuvre may give some indication of valve function. If the valves are clearly seen as thin bicuspid structures it can be assumed that they will function normally.

While any vein in the deep venous system may remain obstructed after thrombosis, the commonest sites are the left common iliac vein, the proximal part of the superficial femoral vein in the adductor canal, and the inferior vena cava.

Thrombotic occlusions at these sites may be continuous producing a long segmental block or there may be short occlusions with normal intervening vein. These occlusions are frequently accompanied by extensive vein damage in the calf and elsewhere.

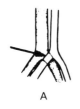

A

Usual site of
compression (80%)

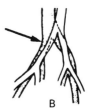

B

Complete overriding
of vena caval
bifurcation by aorta

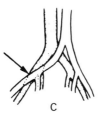

C

Compression of right
external iliac vein in
bifurcation of right
common iliac artery

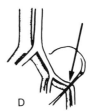

D

Compression at
inguinal ligament

Fig. 7.12 Diagram of sites of possible venous compression by neighbouring structures. (From Cockett et al, 1967)[6]

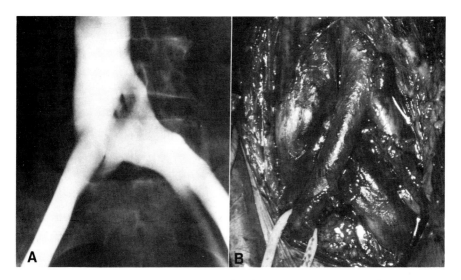

Fig. 7.13 (A) Classical iliac vein compression defect. Note adhesions due to past thrombosis. (B) Operative photograph showing compression of the termination of the left common iliac vein by the right common iliac artery (pulled aside). By kind permission of Mr. F. B. Cockett.

The iliac veins

The predisposition of certain sites to thrombosis is a matter of speculation. In some sites however anatomical structures adjacent to the veins may compress them leading to venous thrombosis and post thrombotic obstruction (Fig. 7.12). The frequency of thrombosis of the left common iliac vein may be related to the fact that this vein is crossed by the right common iliac artery[5,6,20] (Fig. 7.13) as discussed in previous chapters. The higher incidence of thrombosis at the junction of the common femoral vein and the external iliac vein may be due to pressure at this site from the inguinal ligament (Fig. 7.14).

Unlike peripheral deep vein thrombosis which may be symptom free, iliofemoral venous thrombosis usually causes massive swelling of the leg, so called phlegmasia alba dolens. Moreover, whereas peripheral deep veins usually recanalise, the iliac veins often fail to do so and permanent occlusion may result; even if recanalisation does occur, the vein is frequently left narrowed with a long segment of stenosis, and abnormal collateral pathways (Fig. 7.15).

Seventy per cent of stenoses affect the left common iliac vein as distinct from the right common iliac vein because of the higher incidence of thrombosis at this site.

The venous collaterals in external and common iliac vein obstruction have been described in detail by Lea Thomas and his colleagues[16] and are shown in diagrammatic form in Fig. 7.16.

The presence, number, position and direction of flow of contrast medium in these collaterals, demonstrated phlebographically, is important for confirmation of the site and significance of obstruction.

Comparison of patients with peripheral deep venous incompetence and patients who have local iliac vein obstruction, shows that swelling of the affected limb resulting from venous incompetence is usually less important to the patient than the calf pain which develops on exercise in cases of iliac vein obstruction.[23] This bursting pain was first described by Bauer[1] as a post thrombotic symptom and was experienced by patients with obstruction to venous return and not those with peripheral deep vein incompetence alone. Because this pain develops after exercise it has been termed venous claudication.[5,6,22]

The inferior vena cava

Acute venous thrombosis of the inferior vena cava is usually an extension of iliofemoral thrombosis. Because of the large size of the inferior vena cava and its major tributaries, thrombus at these sites is a common source of lethal pulmonary

---

Fig. 7.14 Localised post thrombotic changes in the left common femoral and common iliac veins in the region of the inguinal ligament.

Fig. 7.15 (A) Left femoral vein injection showing localised occlusion of the left common iliac vein. Note the enlarged left obturator vein acting as a collateral. Other collaterals arise from the pre-sacral plexus, both ascending lumbar veins and the vertebral plexus. (B) Another patient with a long stenosis involving the left common femoral vein and the external iliac vein. The main collaterals are through the pubic veins. Other collaterals are through the visceral and presacral plexuses.

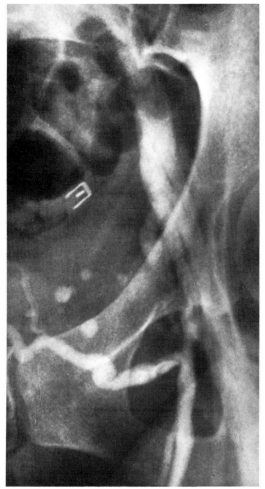

Fig. 7.14

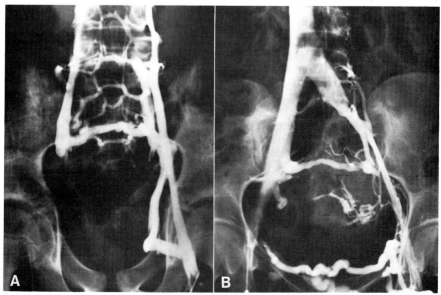

Fig. 7.15

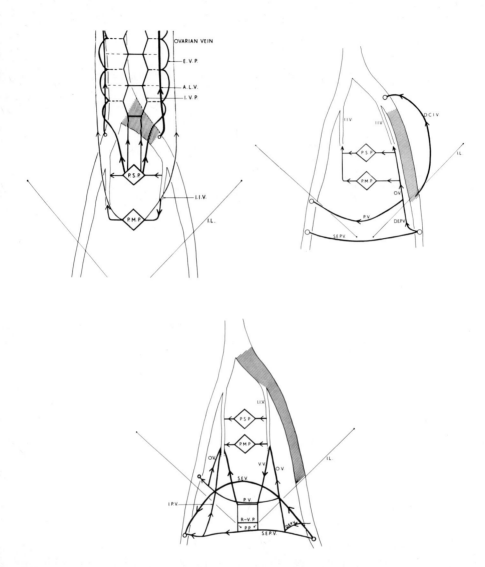

Fig. 7.16 Diagram of the venous collaterals in external and common iliac vein obstruction. The shaded parts represent obstructed segments. The arrows indicate the direction of blood flow.

SEV = superficial epigastric vein
DEPV = deep external pudendal vein
SEPV = superficial external pudendal vein
PP = peroneal plexus
PV = pubic veins
IPV = internal pudendal veins
RVP = rectovesical plexus
PSP = presacral plexus
PMP = parametrial (visceral) plexus
IIV = internal iliac vein
VV = vesical vein
OV = obturator vein
IVP = internal vertebral plexus
EVP = external vertebral plexus
ALV = ascending lumbar vein

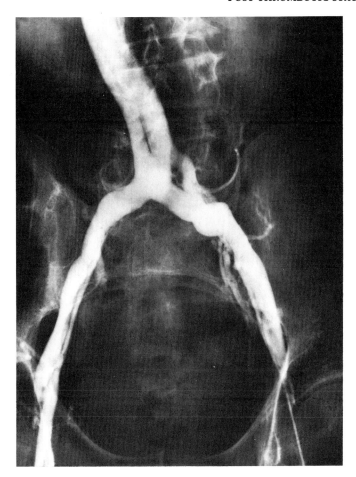

Fig. 7.17 Recanalisation changes are present in the iliac veins and inferior vena cava, indicated by the irregularity of the lumens. There are no collateral veins indicating that the conduit function is adequate. The patient was symptom free.

embolism and the presence of loose thrombus is an indication for thrombectomy with or without plication of the vein above. This aspect has been discussed in Chapter 6. As with the iliac veins, the inferior vena cava following thrombosis frequently does not return to normal, either remaining occluded or narrowed by recanalisation changes.

Inferior vena caval obstruction is rarely an isolated finding being often associated with changes in the iliofemoral segments. This accounts for the clinical findings of dilated superficial abdominal wall collaterals in which the blood flow is towards the head.[16, 21, 26, 28, 29]

Clinically the patients invariably have swollen oedematous legs; venous ulceration and claudication may also occur.

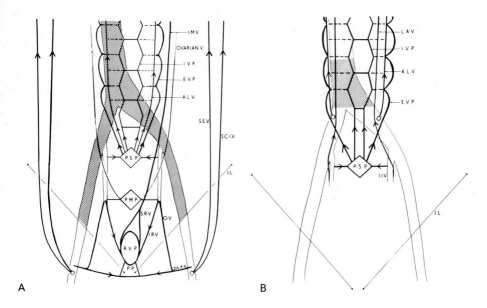

Fig. 7.18 Diagram of collateral circulation in inferior vena caval obstruction. The shaded areas represent the obstructed segments. (A) Combined iliac vein and inferior vena caval obstruction. (B) Localised lower inferior vena caval obstruction.

PP = pudendal plexus
DEPV = deep external pudendal vein
RVP = recto-vesical plexus
IRV = inferior rectal vein
OV = obturator vein
SRV = superior rectal vein
PMP = parametrial (pelvic visceral) plexus
SEV = superficial epigastric vein

SCIV = superficial circumflex iliac vein
PSP = presacral plexus
ALV = ascending lumbar vein
IVP = internal vertebral plexus
EVP = external vertebral plexus
LAV = lumbo azygos vein
IMV = inferior mesenteric vein

Proteinuria may be present if the renal veins are involved but this is not a reliable sign of renal impairment as for many years collateral veins may provide adequate venous drainage for the kidneys.

As the external and common iliac veins and inferior vena cava are devoid of valves, this part of the venous system forms purely a conduit function. Following recanalisation only minimal changes may be present on the phlebogram (Fig. 7.17) and provided this conduit remains adequate no symptoms will arise. The recanalisation changes may leave the inferior vena cava narrowed although not completely occluded. In this case collateral veins will develop indicating that the conduit function is potentially inadequate.

Inferior vena caval occlusion may be associated in some patients with a high aortic bifurcation which results in compression of the cava by the right common iliac artery[15] (see Fig. 7.12B).

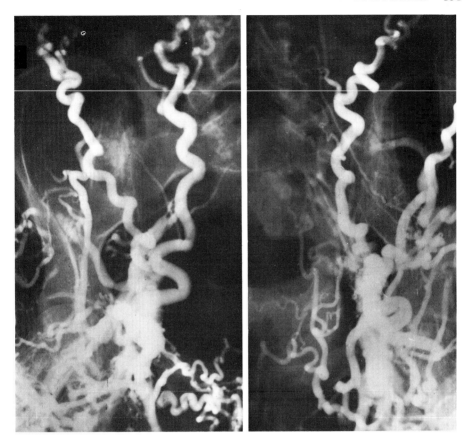

Fig. 7.19 Bilateral pertrochanteric intraosseous phlebogram. The external and common iliac veins are occluded on both sides. The collateral veins bypassing the obstruction are the superficial epigastric (situated more medially) and the superficial circumflex iliac veins. The inferior vena cava has not been demonstrated and is probably occluded. Further studies including retrograde inferior vena cavography would be necessary to be quite certain. These collateral veins should be clinically obvious.

Collateral pathways in inferior vena caval obstruction
The site and size of collateral veins bypassing inferior vena caval obstruction depend on the site of the obstruction and its severity.

The main anterior wall collaterals demonstrated by phlebography are the superficial epigastric and superficial circumflex iliac veins and occasionally the lateral thoracic veins. These collaterals are seen when the obstruction extends into the external iliac veins. If the occlusion is limited to the inferior vena cava the collateral circulation will be primarily through the pudendal, pelvic visceral, and vertebral plexuses draining into the lumbo-azygos system (Fig. 7.18, 7.19, 7.20).

For a detailed description of these collateral pathways the reader is referred to the paper by Fletcher and Lea Thomas.[10]

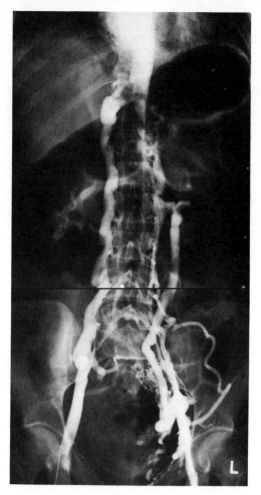

Fig. 7.20 There is total occlusion of both common iliac veins and the inferior vena cava. The collateral veins are the left gonadal vein, which is enlarged, and both lumbo-azygos systems. These collaterals are on the posterior abdominal wall and are not detectable on clinical examination.

VENOUS PRESSURE MEASUREMENTS

These are described here because they can be combined with phlebography using the same venepuncture. Only the principles are outlined here and for a comprehensive account of the techniques and interpretation readers are referred to the monographs by Ludbrook[18] and Van der Hyde.[30] Peripheral venous pressure measurements can give information about a wide variety of venous functions including abnormalities due to venous incompetence, obstruction and malfunction of the muscular venous pumps.

A simple direct pressure measuring device is the saline manometer which, although largely discarded as a research instrument in favour of more sophisticated electronic instruments, is entirely adequate for pressure measurements.

POST THROMBOTIC STATES 133

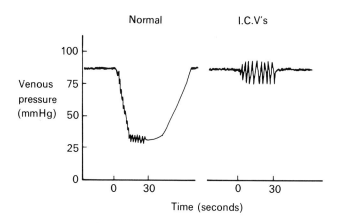

Fig. 7.21 Foot venous pressure tracing. The left hand tracing shows the normal fall in pressure on exercise when the calf muscle pump is fully functional. The right hand tracing is from a patient with incompetent communicating veins (I.C.V's) and in this case the pressure does not fall on exercise.

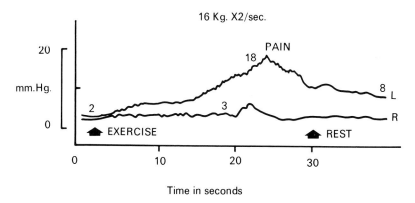

Fig. 7.22 Femoral vein pressure tracing. There is post thrombotic obstruction of the left common iliac vein resulting in a rise in pressure and pain on exercise. (After Negus, 1970)[23].

Disposable presterilised apparatus is now readily available. The pressure is usually expressed in centimetres of water in relation to an anatomical reference point. This may be converted into mmHg by the multiplication factor 0.74.

For peripheral venous pressure measurements the reference point usually employed is the tip of the intravenous cannula or needle. It is important to measure the venous pressures on both sides as the difference between them is more accurate than the absolute pressure on one side. The latter is affected by a number of factors, particularly the inadvertent performance of a Valsalva manoeuvre by the patient.

Foot vein pressures

In normal legs, with the patient erect, the venous pressure in the veins of the foot falls rapidly on exercise to between 0 and 50mm Hg and then on cessation of exercise rises slowly over about 30 seconds to its previous height of 90mm Hg.[7, 27] In patients with primary varicose veins the pressure falls to a value of 44 to 60mm Hg when exercise is stopped.[7]

In deep vein incompetence there is a failure of the foot venous pressure to fall on exercise with an immediate return to the pre-exercise level when the exercise is stopped (Fig. 7.21).

The practical way in which combined foot vein pressures and phlebography can be used in patient management has been suggested by Burnand and his colleagues.[3] They concluded that if a fall in pressure during exercise is increased by the application of a venous occluding thigh tourniquet, and the deep veins appear to be normal on phlebography, then there is a good chance of restoring the superficial pressure to normal by orthodox surgical ligation of the saphenous vein and any incompetent communicating veins. On the other hand if there is no change in the exercising pressure in patients with incompetent communicating veins and the deep veins are abnormal on the phlebogram, then ligation of the incompetent communicating veins is likely to be associated with a higher recurrence of venous ulcers.

Femoral vein pressures

The technique recommended by Negus[23] is to insert a fine bore needle under local anaesthetic into the femoral vein at the groin and connect this either to a strain gauge transducer or a saline manometer. With the patient supine the normal venous pressure is usually between 8 and 12mm Hg. Obstruction to one or other iliac veins produces a rise in pressure above that of the opposite vein and this pressure rise can be amplified by exercising the affected limb.[24]

In patients with unilateral iliac vein obstruction the mean pressure is usually about 5mm Hg above that of the normal limb and on exercise rises to about 10mm Hg. In normal limbs exercise produces a rise in venous pressure of less than 2mm Hg (Fig. 7.22).

Thus phlebography and iliac vein pressure measurements are of great value in both diagnosis and in evaluating the prognosis of post thrombotic iliac vein obstruction.

RECONSTRUCTIVE SURGERY IN ILIAC VEIN OBSTRUCTION

The principle of these bypass operations, which have now become a practical possibility as a result or the work of Palma[25] is to divide a saphenous vein on the healthy side at the level of the knee, free it from all collateral vessels to its junction with the femoral vein, and transpose it in a subcutaneous tunnel across the pubis to the diseased side where it is anastamosed with the femoral vein (Fig. 7.23). The shunt may be kept open by a temporary arteriovenous fistula.

The long term benefits of this therapy have yet to be evaluated, but if patient selection is limited to those with diseased pelvic veins and where the distal leg

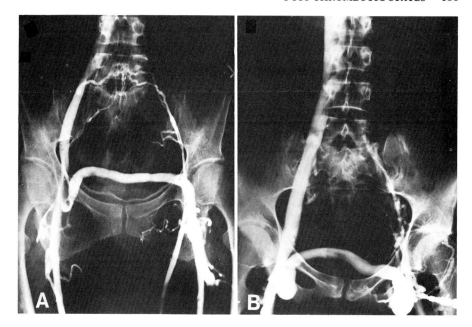

Fig. 7.23 Two examples of Palma's operation. In both instances there is a severe obstruction to the common and external iliac veins. The right long saphenous vein has been transposed and joined to the left femoral vein in order to provide venous drainage of the left lower limb. (A) An example 5 years after the operation with the patent bypass demonstrated by an ascending phlebogram. (B) An example 12 years after surgery in which the patent bypass has been shown by direct left femoral vein injection (Radiographs by courtesy of Dr Robert May).

veins are well preserved, reasonably good symptomatic results may be expected[19].

To demonstrate the patency of the bypass graft an ascending phlebogram of the affected leg is carried out when contrast is shown to drain from the affected leg into the opposite common femoral vein. Alternatively the vein below the bypass may be injected directly.

REFERENCES

1. Bauer G A 1948 The aetiology of leg ulcers and their treatment by resection of popliteal vein. J Int Chir 8: 937
2. Bauer G 1950 Division of popliteal vein in the treatment of so called varicose ulceration. Brit Med J 2: 318
3. Burnand K G, O'Donnell T F, Lea Thomas M, Browse N L 1977 The relative importance of incompetent communicating veins in the production of varicose veins and venous ulcers. Surgery 82: 9
4. Boyd A M, Jepson R P, Ratcliffe A H, Rose S S 1952 Logical management of chronic ulcers of the leg. Angiology 3: 207
5. Cockett F B, Lea Thomas M 1965 The iliac compression syndrome. Brit J Surg 52: 816
6. Cockett F B, Lea Thomas M, Negus D 1967 Iliac vein compression – Its relation to iliofemoral thrombosis and the post thrombotic syndrome. Brit Med J 2: 14

7. Dodd H, Cockett F B 1976 Pathology and surgery of the veins of the lower limb 2nd edn Churchill Livingstone, Edinburgh
8. Edwards E A, Edwards J E 1937 The effect of thrombophlebitis on the venous valve. Surg Gynec Obstet 65: 310
9. Field E S, Kakkar V V, Stephenson G, Nicolaides A N 1972 The value of cine phlebography in detecting incompetent venous valves in the post phlebitic state. Brit J Surg 59: 304
10. Fletcher E W L, Lea Thomas M 1968 Chronic post-thrombotic obstruction of the inferior vena cava investigation by cavography. A report of two cases. Am J. Roentgenol 102: 363
11. Gay J 1868 On varicose diseases of the lower extremities. The Lettsomian Lectures of 1967. Churchill, London
12. Halliday P 1968 Phlebography of the lower limb. Brit J Surg 55: 220
13. Homans J 1916 The operative treatment of varicose veins and ulcers based upon a classification of these lesions. Surg Gynec Obstet 22: 143
14. Kakkar V V, Howe C T, Laws J W, Flanc C 1969 Late results of deep vein thrombosis. Brit Med J 1: 810
15. Lea Thomas M, Andress M R 1970 Phlebographic changes in arteriomegaly. Acta Radiol Diag 10: 427
16. Lea Thomas M, Fletcher E W L, Cockett F B, Negus N 1967 Venous collaterals in external and iliac vein obstruction. Clin Radiol 18: 402
17. Linton R R 1952 Modern concepts in the treatment of post phlebitic syndrome with ulceration of the lower extremity. Angiology 3: 431
18. Ludbrook J 1972 The analysis of the venous system. Hans Huber Publishers, Bern
19. Hay R, De Weese J A 1979 In: (ed) May R Surgery of the veins of the leg and pelvis. Georg Thieme Publishers, Stuttgart
20. May R, Thurner J 1956 Ein gefassporn in der Vena iliaca communis sinistra als Ursache der uberwiegend linkseitigen Beckenvenen-thrombosen Z Kreisl-Forsch 45: 912
21. Missal M E, Robinson J A, Tatum R W 1965 Inferior vena caval obstruction; Clinical manifestations, diagnostic methods and radiological problems. Ann Int Med 62: 133
22. Negus D 1968 Calf pain in the post thrombotic syndrome. Brit Med J 2: 156
23. Negus D 1970. The post thrombotic syndrome. Ann Roy Coll Surg Egl 47: 92
24. Negus D, Cockett F B 1967 Femoral vein pressures in post phlebitic iliac vein obstruction. Brit J Surg 54: 522
25. Palma E C 1960 Vein transplants and grafts in the surgical treatment of the post phlebitic syndrome. J Cardiovasc Surg 1: 94
26. Pleasants J H 1911 Obstruction of the inferior vena cava with a report of 18 cases. John Hopkins Hosp Rep 16: 363
27. Pollack A A, Wood E N 1949 Venous pressure in the saphenous vein at the ankle in man during exercise and changes in posture. J Appl Physiol 9: 649
28. Robinson L S 1949 Collateral circulation following investigation of the inferior vena cava. Surgery 25: 329
29. Stein S, Blumsohn D 1962 Clinical and radiological observations of inferior vena caval obstruction. Brit J Radiol 35: 159
30. Van Der Hyde M N 1961 Phlebography and venous pressure delineation. H E Stenfert Kroese N V, Leiden
31. Widmer L K 1967 Venenerkron bei 1800 Berufstiagen, Basler. Studie Schweie Med Wschr 97: 107

8

Varicose Veins

INTRODUCTION

A vein is considered to be varicose when it is dilated, elongated, tortuous and has incompetent valves.

Varicose veins are common, although the exact incidence remains unclear despite the detailed work of Borschberg[1] and Widmer.[9] This uncertainty presumably results from the differing views of classification and definition but an incidence of between 10 per cent and 20 per cent in the western world has been suggested.[3]

It is convenient to classify the aetiology of varicose veins into primary and secondary.

PRIMARY VARICOSE VEINS

In primary varicose veins by definition the deep venous system is normal, and the valves in the deep veins and communicating veins are competent. The superficial veins, mainly the long and short saphenous systems have incompetent valves at or near their origins allowing increased hydrostatic pressure due to the upright posture to be transmitted to these veins.

The cause of primary varicose veins is unknown although a number of aetiological factors have been suggested, including genetic, familial and racial, also diet, obesity and pregnancy.

Phlebography in primary varicose veins

Phlebography plays little part in the diagnosis and management of primary varicose veins and the diagnosis is usually made clinically, combined where necessary with simple non-invasive tests. In certain situations however phlebography has a diagnostic role. Examples include obese patients where the clinical tests are difficult to interpret; when varicose veins are in an unusual site; if unexpected recurrences occur after treatment, and in young patients where there is a possibility of an underlying congenital anomaly affecting the venous system. Perhaps the commonest reason for requesting phlebography is where a patient is thought to have primary varicose veins, but where there is doubt as to whether the condition is in fact primary or secondary or of mixed aetiology.

138 PHLEBOGRAPHY OF THE LOWER LIMB

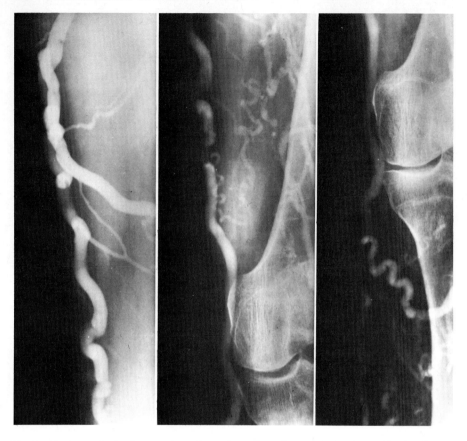

Fig. 8.1 Varicography has been performed in order to show the course of an unusually placed varicose vein. This example is an uncommon congenital anomaly in a young man. The deep venous system (not demonstrated by this technique) was normal.

Direct injection of a varicose vein itself, so called varicography, will determine the route of drainage of the varicosity through the communicating veins in the deep venous system and give an idea of the extent of the varicosities (Figs. 8.1 and 8.2).

Ascending phlebography by injection of superficial veins of the ankle without the use of tourniquets and the patient in the horizontal position with a slight head down tilt will also demonstrate the extent of varicose veins (Fig. 8.3). When there is clinical doubt about the diagnosis of primary varicose veins it is necessary to demonstrate the normality or otherwise of the deep venous system and the presence or absence of incompetent communicating veins. The phlebographic technique used is the standard one for ascending phlebography with a steep table tilt downwards and the use of tourniquets as described in Chapter 3 (Fig. 8.4).

VARICOSE VEINS 139

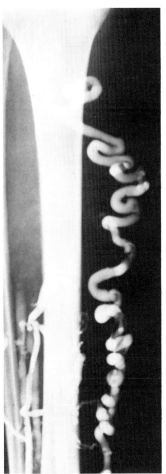

Fig. 8.2 Long saphenous varicosity shown by varicography. Injection of a varicose vein at the ankle, without an ankle tourniquet. A little deep venous filling is present in the lower calf and the deep veins appear normal.

Fig. 8.2

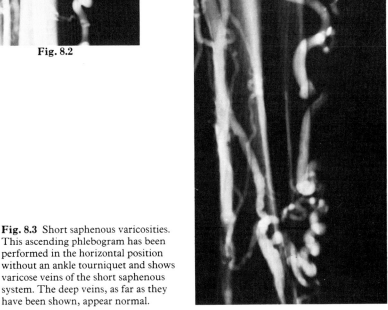

Fig. 8.3 Short saphenous varicosities. This ascending phlebogram has been performed in the horizontal position without an ankle tourniquet and shows varicose veins of the short saphenous system. The deep veins, as far as they have been shown, appear normal.

Fig. 8.3

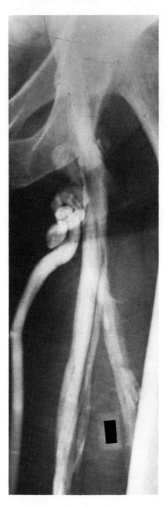

Fig. 8.4 Mixed primary and secondary varicose veins. There is incompetence of the long saphenous vein with proximal tortuosity. There are also recanalisation changes in the femoral vein. A Valsalva manoeuvre has shown these abnormalities more clearly.

Incompetence of the short saphenous vein can be established if a Valsalva manoeuvre is performed when the popliteal vein is filled with contrast medium (Fig. 8.5); similarly a Valsalva manoeuvre when the common femoral vein is filled will demonstrate incompetence of the long saphenous vein (see Fig. 8.4). Another way to show short saphenous vein incompetence is to place a tourniquet above its termination so that the incompetent vein fills in a retrograde manner (Fig. 8.6).

A 'saphena-varix' frequently accompanies long saphenous vein incompetence and can be demonstrated by ascending phlebography with a Valsalva manoeuvre (Fig. 8.7).

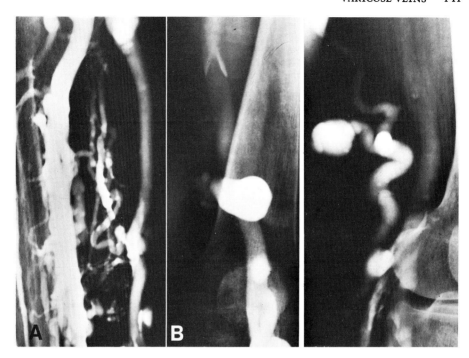

Fig. 8.5 Examples of the use of the Valsalva manoeuvre to show short saphenous vein incompetence. (A) Lateral view of calf. The short saphenous vein lies posteriorly and is dilated and valveless. The deep veins are normal. (B) Another patient. (Frontal and lateral projections.) The deep veins are normal. The Valsalva manoeuvre impedes central flow in the femoral vein so that the varicose short saphenous vein is better demonstrated.

In the treatment of short saphenous vein incompetence some surgeons prefer to ligate the vein at its junction with the popliteal vein rather than where it pierces the deep fascia.[5] The rather variable termination of the short saphenous vein can be shown on a lateral projection of the popliteal region if ascending phlebography is performed without an ankle tourniquet, so that the superficial venous system is not occluded (Fig. 8.8). This problem does not arise at the termination of the long saphenous vein which has a relatively constant position and can be identified readily from the conventional surgical incision.

Descending phlebography can also be used to demonstrate superficial vein incompetence, either by puncturing the common femoral vein or the popliteal vein and taking radiographs in the upright posture. This technique is rarely required in practice however.

SECONDARY VARICOSE VEINS

Secondary varicose veins are due to some abnormality of the deep venous system, usually post thrombotic destruction of the valves, incompetent communicating veins, or in association with congenital venous or arteriovenous anomalies.

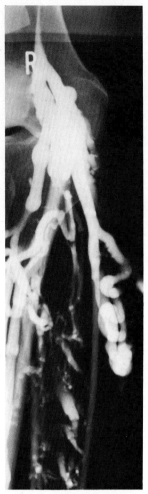
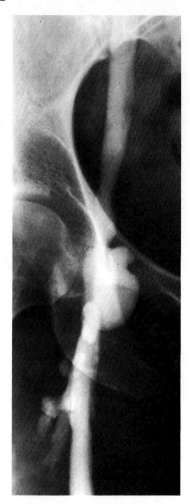

Fig. 8.6   **Fig. 8.7**

Fig. 8.6 Short saphenous vein varicosities (lateral view). A tourniquet is placed above the termination of the vein to impede the escape of contrast medium from the calf veins.

Fig. 8.7 An example of a 'saphena-varix' demonstrated by ascending phlebography with a Valsalva manoeuvre.

Phlebography in secondary varicose veins

The various techniques which may be used to demonstrate the deep venous system have already been discussed in detail in Chapter 3 and referred to again in Chapter 7.

As incompetent communicating veins are the most important feature of secondary varicose veins, some of the more significant aspects of their demonstration and appearances are described again here.

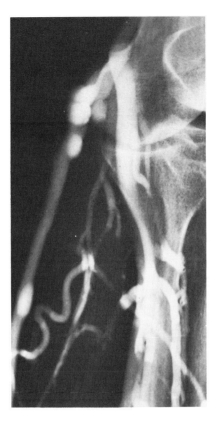

Fig. 8.8 The termination of the short saphenous vein shown in a lateral projection of an ascending phlebogram of the calf and popliteal region.

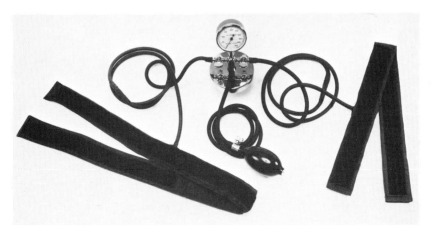

Fig. 8.9 Pneumatic cuffs for ascending phlebography to demonstrate incompetence in the communicating veins.

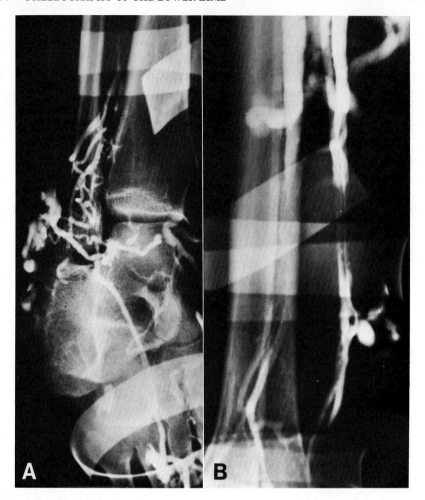

Fig. 8.10 The use of multiple tourniquets. (A) An infra-malleolar incompetent communicating vein is shown by using a tourniquet around the forefoot and also one above the ankle. (B) An incompetent communicating vein arising from the posterior tibial vein is shown in this lateral projection of the calf. An additional tourniquet has been placed above the ankle tourniquet to prevent superficial venous filling which might obscure incompetent communicating veins higher up. This incompetent communicating vein is also better shown because the higher tourniquet slightly obstructs the deep venous system assisting contrast to enter it.

Incompetent communicating veins are readily recognised on ascending phlebography.[6] The standard technique of ascending phlebography is modified by applying ankle tourniquets sufficiently tightly to occlude the superficial veins thus allowing the radiologist to identify by fluoroscopy any reverse flow from the deep to the superficial veins through the incompetent communicating veins. The above knee tourniquet should also be applied more tightly so that more contrast medium is directed into any incompetent communicating veins.

VARICOSE VEINS 145

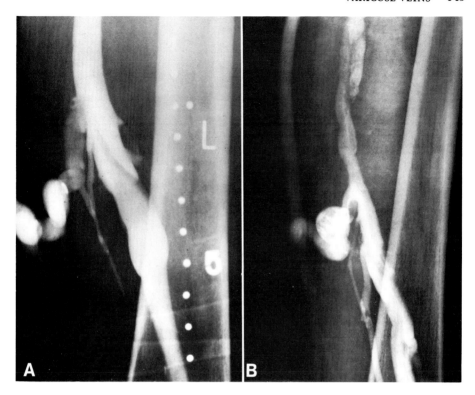

Fig. 8.11 A Hunter's canal incompetent communicating vein shown by a Valsalva manoeuvre and a high tourniquet (not shown). (A) The large incompetent communicating vein is well demonstrated. The femoral vein appears normal. The metallic markers facilitate accurate localisation. (B) In this example there are severe recanalisation changes in the femoral vein. The incompetent communicating vein can be seen entering the long saphenous vein. Note that the long saphenous vein above the incompetent communicating vein is denser than that below confirming the site of entry of the contrast.

The self-fastening rubber ankle tourniquet used in the standard technique of ascending phlebography is often adequate if applied tightly enough, and its tightness checked by a test injection of contrast medium at the start of the examination. Pneumatic cuffs described by Craig[2] permit better control of the occlusion of the superficial veins at the ankle. The cuffs are attached to a manometer so that the pressure can be measured and the pressure of each cuff can be separately controlled (Fig. 8.9). In most instances the lower cuff is placed just above the ankle, but if a low, infra-malleolar, incompetent communicating vein is suspected or there is ankle ulceration, the cuff can be placed around the forefoot. The upper cuff should be positioned just above the adductor hiatus. The ankle cuff is inflated to 120mm Hg and the mid thigh cuff to about 200mm Hg. These pressures do not occlude the arterial flow because the cuffs are narrow, but the

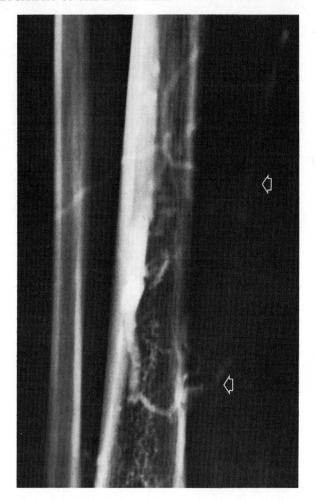

Fig. 8.12 Ascending phlebogram by the intraosseous technique using the medial malleolus. This shows two incompetent communicating veins (arrows).

pressures need to be varied according to the size of the patient's leg and their effectiveness checked by fluoroscopy.[7] On no account should an ankle cuff be excessively tight otherwise veins distal to the cuff may rupture with extravasation of the contrast medium. If selective deep venous filling cannot be obtained by using a comfortable pressure and a steep table tilt, the examination should be abandoned and the intraosseous route used instead (see Ch. 4).

When the superficial veins are opacified inadvertently, or via an incompetent communicating vein, the veins should be cleared of contrast medium by injecting physiological saline and a second more proximal tourniquet applied before proceeding with the examination. Occasionally it may be necessary to apply a series of tourniquets to occlude incompetent communicating veins which may

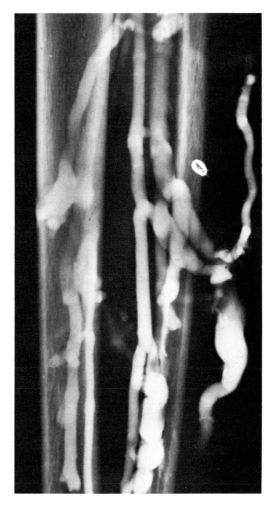

Fig. 8.13 Calf phlebogram showing two large incompetent communicating veins with the typical features of dilatation, absence of valves, tortuosity, and a tendency to run downwards initially.

otherwise allow 'flooding' of the superficial venous system by contrast, thus obscuring more proximal incompetent veins (Fig. 8.10).

As a rule, straight films (postero-anterior) only are required. Lateral films are often difficult to interpret, particularly in the presence of superficial venous filling. The origins of nearly all the clinically important communicating veins connecting the posterior tibial and peroneal veins with the superficial veins are shown the postero-anterior projection. Oblique views occasionally may be necessary to distinguish a superficial from a deep vein. The former move in a wider arc during rotation of the leg than the latter which are closely related to the tibia and fibula.

Fig. 8.14 Thrombus in a communicating vein. Destruction of the valves by thrombus renders the vein permanently incompetent.

Incompetent communicating veins in the thigh can sometimes only be demonstrated by a Valsalva manoeuvre (Fig. 8.11).

Accuracy of phlebography in demonstrating incompetent communicating veins

The accuracy of ascending phlebography in the detection of incompetent communicating veins varies in different series but in the author's experience[6] 81 per cent of incompetent communicating veins were subsequently confirmed at operation.

An intraosseous injection into a malleolus or the os calcis is probably the most accurate method of demonstrating incompetent communicating veins[8] as with this technique contrast passes preferentially into the deep veins[4] (Fig. 8.12). The details of the technique are described in Chapter 3.

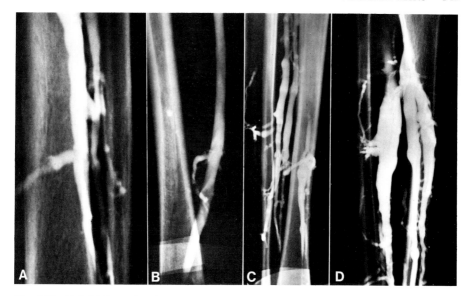

Fig. 8.15 (A & B) Examples of normal communicating veins. Normal valves can be seen. Retrograde flow of a little contrast medium through valves should not be regarded as a sign of permanent incompetence. (C & D) The lateral communicating veins shown here do not connect with varicose veins and should not therefore be regarded as pathological.

Appearances of incompetent communicating veins

In addition to the retrograde flow of contrast medium in incompetent communicating veins, these veins lack normal valves, are abnormally wide and tortuous, and are dilated at their distal ends where they join a varicose vein. They tend to run downwards initially before passing upwards to pierce the deep fascia (Fig. 8.13). It is thought that communicating veins are rendered incompetent by thrombosis which destroys their valves (Fig. 8.14). Communicating veins without these additional features should be considered normal, as occasionally a normally functioning communicating vein will allow contrast medium to pass from the deep to the superficial venous system (Fig. 8.15).

Accurate localisation of incompetent communicating veins is possible by means of a plastic ruler with embedded ball bearings at 1cm intervals, placed beneath the leg during phlebography. The communicating veins can be measured in relation to the patient's leg using the tips of the medial or lateral malleolus as landmarks (Fig. 8.16).

Defining the exact position of incompetent communicating veins can sometimes be difficult when varicose veins fill rapidly through a distal incompetent communicating vein thus obscuring others which may be present. To avoid this early films should be taken so that the *origin* of the incompetent communicating veins from the deep venous system can be clearly seen (Fig. 8.17).

150 PHLEBOGRAPHY OF THE LOWER LIMB

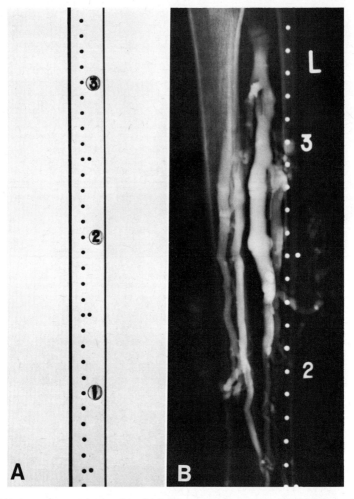

Fig. 8.16 (A) A transparent plastic ruler with ball bearings set at 1cm intervals which is placed beneath the leg during ascending phlebography to aid localisation of incompetent communicating veins. (B) This phlebogram shows the ruler in position. The site of the incompetent communicating veins can thus be related to bony land marks, such as the malleoli, on the patient's leg.

Generally it can be assumed that medial incompetent communicating veins arise from the posterior tibial veins and lateral communicating veins from the peroneal veins. Clinically important communicating veins do not often arise from the anterior tibial veins. Since incompetent communicating veins pass outwards from the stem veins to pierce the deep fascia it is this area that should be demonstrated phlebographically in order to localise the precise level. If the exact site is not clearly demonstrated on the phlebogram the examination is of little

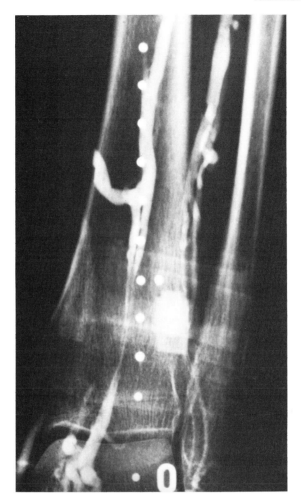

Fig. 8.17 The origin of this medial incompetent communicating vein is clearly shown on this early film.

practical use. Another feature which may be helpful in identifying incompetent communicating veins arises from the observation that the contrast medium is denser in the superficial veins above an incompetent communicating vein than below it (see Fig. 8.11).

If intraosseous phlebography is required, a bone around the ankle remote from any ulceration should be chosen for puncture to avoid the risk of subsequent osteomyelitis.

The importance of phlebography in identifying incompetent communicating veins results from the fact that while most of these veins are relatively constant in position and can be recognised clinically and at surgery (see Ch. 2) some are inconstant in position. The accurate identification and localisation of these inconstantly placed veins thus enables successful treatment to be undertaken and is one of the reasons why phlebography may be useful in the management of varicose veins. In addition phlebography confirms the patency of the deep venous system, if there is any clinical doubt, before stripping of superficial varicose veins.

REFERENCES

1. Borschberg E 1967 Prevalence of varicose veins in the lower extremity. Karger, Basel
2. Craig J O 1964 In (ed) Saxton H M, Strickland B Practical procedures in diagnostic radiology. H K Lewis, London
3. Dodd H, Cockett F B 1976 The pathology and surgery of veins of the lower limb 2nd edn Churchill Livingstone, Edinburgh
4. Halliday P 1968 Phlebography of the lower limb. Brit J Surg 55: 220
5. Hobbs J T 1980 Peroperative venography to ensure accurate sapheno-popliteal vein ligation. Brit Med J 280: 1578
6. Lea Thomas M, McAllister V, Rose D J, Tonge K 1972 A simplified technique of phlebography for the localisation of incompetent perforating veins of the legs. Clin Radiol 23: 86
7. Nicolaides A N, Kakkar V V, Field E S, Renney J T G 1971 The origin of deep vein thrombosis: a venographic study. Brit J Radiol 44: 653
8. Townsend J, Jones H, Williams J 1967 Detection of incompetent perforating veins by venography at operation. Brit Med J 3: 583
9. Widmer L K 1978 Peripheral venous disorders. Hans Huber Publishers, Bern

9

Malformations

INTRODUCTION

Malformations or dysplasias of the venous system are relatively common because of the complicated embryological anatomy. These venous dysplasias range from insignificant anatomical variations to complex angiodysplasias.

There have been many attempts to classify venous malformations, either by syndromes[3,12] or by more comprehensive subdivisions based on haemodynamic, embryologic, pathologic and angiographic concepts.[5,9,16]

No classification has proved entirely satisfactory because as Kinmonth and his colleagues[2] have pointed out, vascular angiodysplasias are frequently mixed, with elements from the arterial, venous and lymphatic systems together with other tissues present within the same lesion. It is probably better to use simple descriptive adjectives to indicate which of the vascular elements predominates as it is this aspect that will affect the treatment.

In this chapter we are only concerned with those malformations which are predominantly venous.

PHLEBOGRAPHY

As the state of the deep venous system plays such an important part in the management of these malformations a meticulous technique of ascending and other forms of phlebography is required. Tight ankle tourniquets are needed to occlude the superficial veins in order to direct the contrast into the deep venous system, otherwise erroneous diagnosis of absent deep veins can easily be made. If the percutaneous technique is not successful, intraosseous phlebography will probably be required.

It is sometimes the best method of demonstrating venous angiomas by injecting into the bone marrow at a site close to the lesion (Fig. 9.1).

In superficial and localised venous dysplasias direct injection into the angioma with a fine gauge needle is often the simplest way of showing the extent of the lesion and its ramifications before surgical removal (Fig. 9.2). In the more extensive angiodysplasias injection into an artery with follow through to the venous phase (arteriophlebography) may be the only way of showing the extent of the abnormality (Figures 9.3 and 9.4). Subtraction films to enhance the image may be used for improved definition (Fig. 9.5).

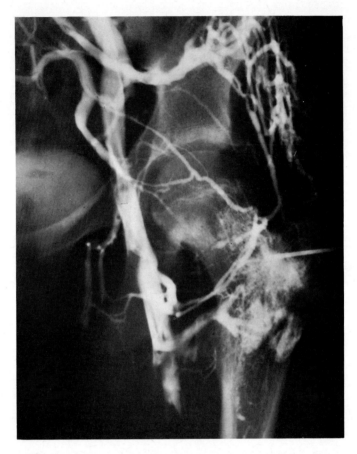

Fig. 9.1 A deep venous angioma of the buttock displayed by pertrochanteric intraosseous phlebography.

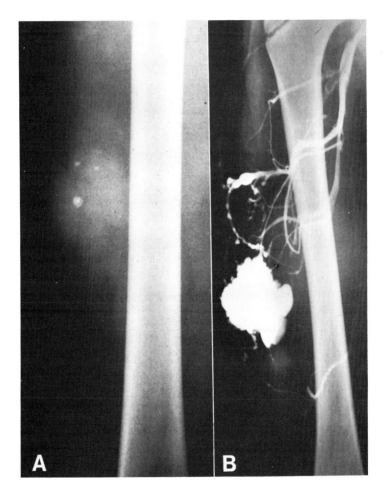

Fig. 9.2 (A) Plain film of the thigh showing phleboliths in a venous angioma. (B) Direct injection of contrast medium into the angioma indicates the size of the lesion and shows that its main connection is with the profunda femoris vein.

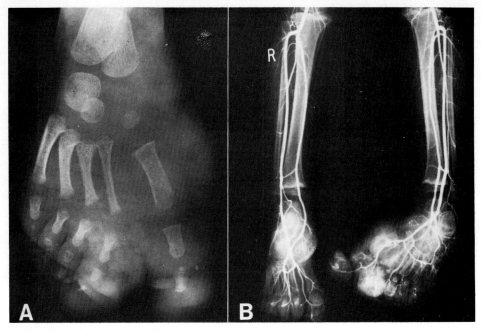

Fig. 9.3 Arteriophlebogram to demonstrate an angioma of the left foot. The arteries to the left foot are not enlarged indicating that this is not an arterio-venous malformation. (A) Plain film showing soft tissue changes in the foot. (B) Arterio phlebogram (intermediate phase) showing the venous malformation already appearing in the left foot.

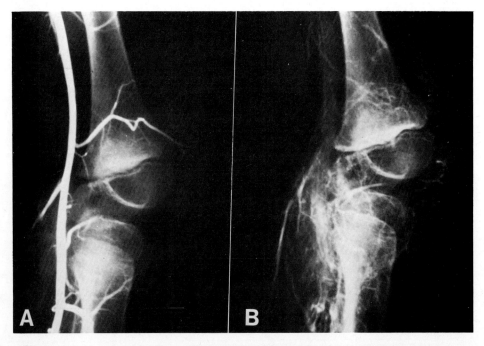

Fig. 9.4 Arteriography with follow through (arteriophlebogram) to show an extensive venous angioma. (A) Arterial phase. The femoral and popliteal arteries are of normal size. A few small tributaries supply the angioma. (B) Venous phase. This shows that the angioma is mainly situated in the popliteal fossa.

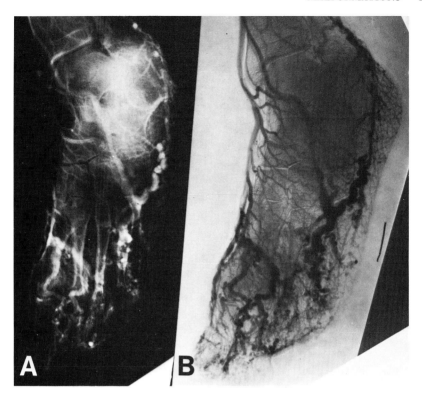

Fig. 9.5 Arteriophlebogram in a young patient with an extensive venous malformation of the foot. There are dilated tortuous veins throughout the forefoot. These vessels could not be demonstrated adequately by direct phlebography. (A) Venous phase of the arteriogram. (B) A subtraction film of the same phase showing the malformation more clearly.

If there is doubt about the presence or absence of an arteriovenous fistula, rapid sequence angiography (Fig. 9.6) will exclude haemodynamically significant fistulae and will demonstrate feeding vessels which may have to be surgically ligated or embolised.

VENOUS DYSPLASIAS

Anatomical variations

The inferior vena cava
The inferior vena cava is a complex structure. Its infra renal part is derived from the posterior cardinal and supra-cardinal veins. The persistence of parts of these embryological trunks may give rise to a variety of anomalies and it has been estimated that about 1 per cent of otherwise normal subjects have such anomalies of the inferior vena cava and its tributaries.[10]

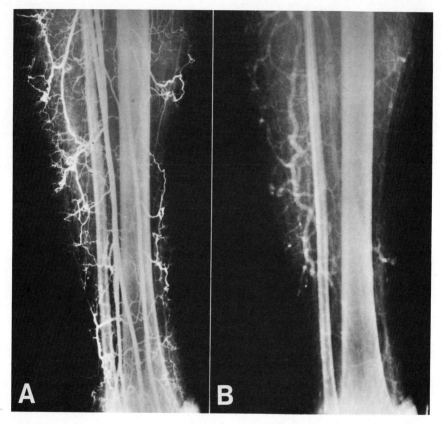

Fig. 9.6 Arteriogram of the lower leg in a patient with arteriovenous fistulae. (A) Arterial phase showing numerous small arteriovenous fistulae throughout the leg. (B) Venous phase.

These anomalies are not discussed here in detail as they are frequently more the concern of the cardiologist than the phlebologist but examples include reduplication (Fig. 9.7), left sided inferior vena cava (Fig. 9.8) and occasionally hypoplasia (Fig. 9.9) or complete absence of the inferior vena cava with compensatory enlargement of the azygos and hemi-azygos systems. Another anomaly, although infrequent, is that of a retrocaval or retroiliac ureter due to the persistence of the right posterior cardinal vein. The distal ureter lies dorsal to the inferior vena cava and is usually recognised on excretion urography or retrograde ureterograms where the ureter is demonstrated medial to its expected anatomical

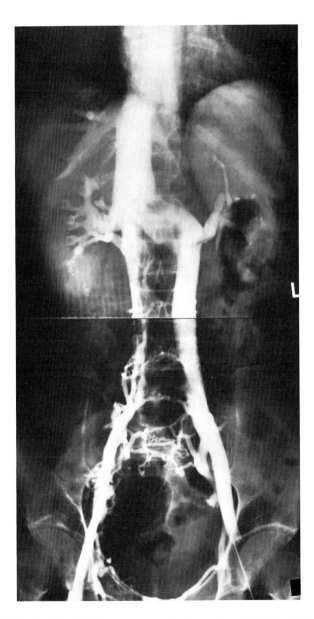

Fig. 9.7 Reduplication of the inferior vena cava. The plexus of vessels to the right of the spine are the remnants of a full sized right sided vessel which has become thrombosed following plication for pulmonary embolism. A phlebogram was not carried out preoperatively and the patient had a further pulmonary embolus through the left trunk.

160 PHLEBOGRAPHY OF THE LOWER LIMB

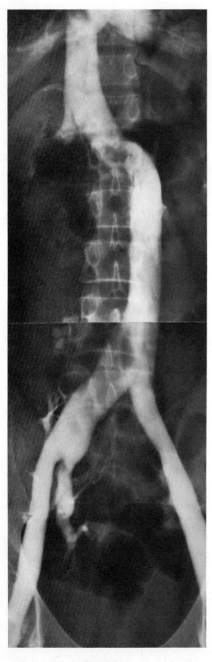

Fig. 9.8

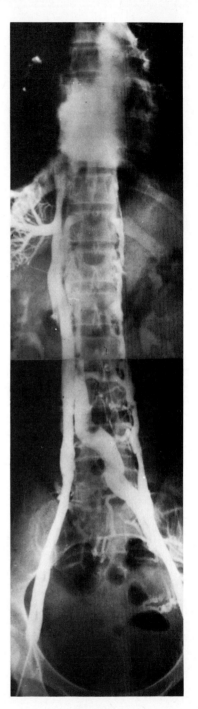

Fig. 9.9

Fig. 9.8 A pure left sided inferior vena cava. The vein crosses to the right to continue cranially as an otherwise normal inferior vena cava.
Fig. 9.9 Congenital hypoplasia of the inferior vena cava.

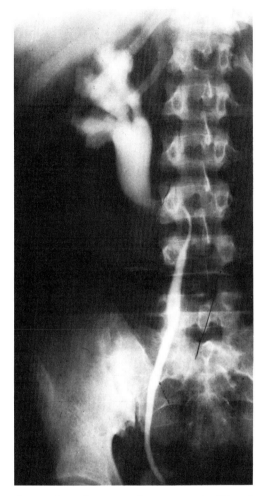

Fig. 9.10 A retrocaval ureter shown by retrograde ureterography. The calyces, pelvis and upper part of the ureter are dilated. Below the obstruction the ureter is displaced towards the mid line.

position. If obstructed, only the pelvi-calyceal system and the proximal third of the ureter is dilated (Fig. 9.10). Inferior vena cavography is the definitive means of establishing the diagnosis but as treatment consists of replacing the obstructed ureter anterior to the cava, this investigation is rarely required before surgery. A left retrocaval ureter may be present with situs inversus.

Many caval abnormalities are not important in themselves, but the possibility of a reduplicated inferior vena cava needs to be borne in mind when plication is being carried out for recurrent pulmonary embolism (see Fig. 9.7).

Membranous stenosis or occlusion of the diaphragmatic part of the inferior vena cava can occur and requires surgical correction.

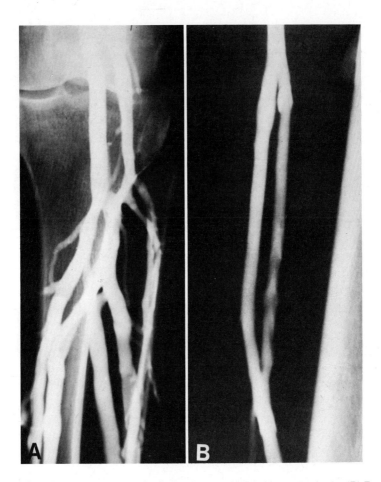

Fig. 9.11 Examples of reduplication of the deep veins. (A) Double popliteal vein. (B) Double femoral vein.

Simple anomalies of the leg veins

These variations have no haemodynamic importance but it is helpful to know about them before undertaking surgery for varicose veins, in the assessment of recurrent varicose veins and when obtaining venous grafts. A knowledge of them is also required so that thay are not confused with significant venous abnormalities.

Numerical anomalies

These are extremely common and have been referred to in Chapter 2. They can

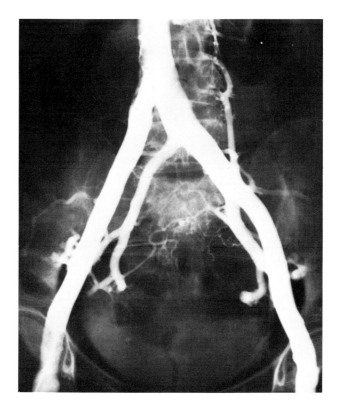

Fig. 9.12 An example of an abnormal termination of a deep vein. The right internal iliac vein joins the left common iliac vein.

occur either in the deep veins, such as the femoral and popliteal veins (Fig. 9.11) or in the superficial veins particularly the long saphenous system.

Anomalies of the termination of the leg veins
These also affect the deep (Fig. 9.12) and superficial veins. The variation in the site of termination of the short saphenous vein may give rise to difficulties if surgical ligation is contemplated, and phlebography beforehand may be required to show the site of the termination (Fig. 9.13).

164 PHLEBOGRAPHY OF THE LOWER LIMB

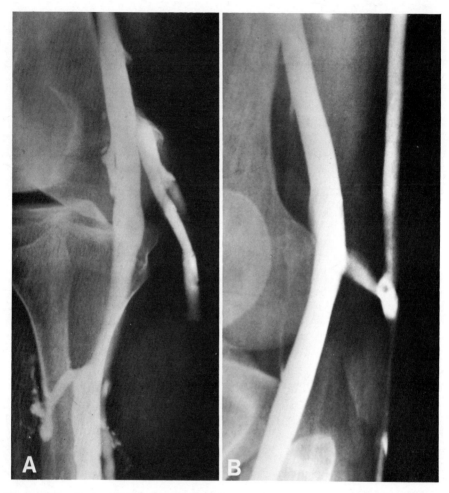

Fig. 9.13 Termination of the short saphenous vein. (A) The commonest site, joining the popliteal vein. (B) In this example there is a connection between the short saphenous vein and the popliteal vein but the short saphenous vein extends upwards into the thigh. This is important if surgical ligation is being considered.

Ectasia

Ectatic veins may be either deep or superficial. Deep vein ectasia frequently affects the soleal muscle sinusoidal veins (Fig. 9.14), and is thought to contribute to thrombosis as a consequence of stasis.

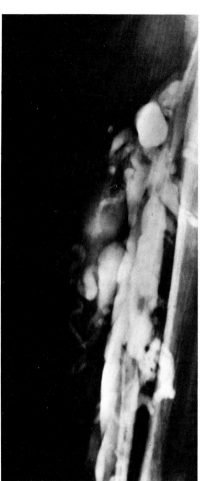

Fig. 9.14

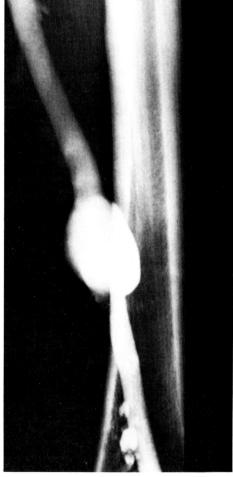

Fig. 9.14 Ectatic calf muscle sinusoidal veins which are prone to thrombosis.

Fig. 9.15 An aneurysm of the superficial femoral vein occurring in the adductor canal.

Fig. 9.15

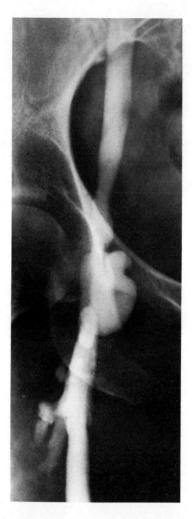

Fig. 9.16 A saphena varix occurring at the termination of the long saphenous vein. This is the commonest site for aneurysmal dilatation.

Venous aneurysms

Aneurysms can occur at any site in the venous system (Fig. 9.15) but the commonest sites are the termination of the long and short saphenous veins-saphena varices, (Figs. 9.16 and 9.17). Occasionally a venous aneurysm may contain thrombus which may embolise.

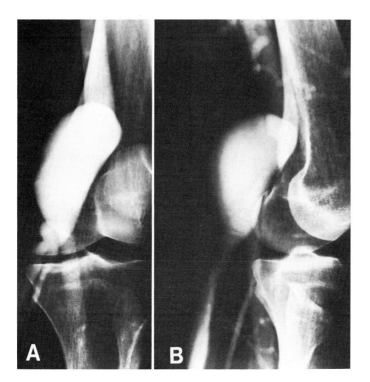

Fig. 9.17 A saphena varix occurring at the termination of the short saphenous vein. (A) Straight projection. (B) Lateral projection.

Aplasia of venous valves

The congenital absence of venous valves is a very rare condition. Such agenesis may affect the whole of the venous system of an extremity or only part of it, or it may be bilateral and symmetrical. The clinical effect is the same as venous insufficiency in which the valves have been destroyed by thrombosis. The symptoms, particularly oedema, occur much earlier than with primary varicose veins, presenting usually at puberty.

The diagnosis can be confirmed by phlebography. Some information can be obtained from an ascending phlebogram including a Valsalva manoeuvre but the most accurate method is by descending phlebography.[13]

168 PHLEBOGRAPHY OF THE LOWER LIMB

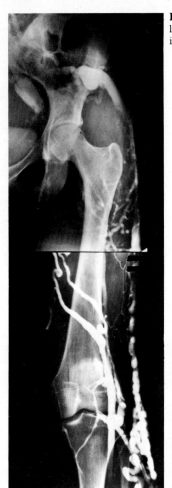

Fig. 9.18

Fig. 9.18 Ascending phlebogram of leg showing the classical lateral venous channel, which in this case joins the left common iliac vein, in a patient with Klippel-Trenaunay syndrome.

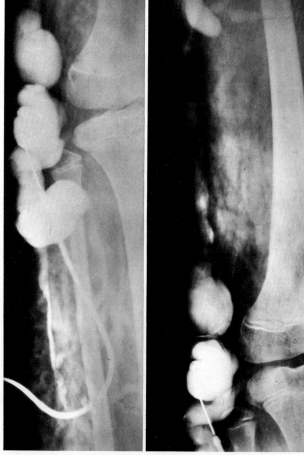

Fig. 9.19 Direct injection into the lateral venous channel in a patient with Klippel-Trenaunay syndrome. The channel is dilated, tortuous and valveless. Direct injection is often a useful way to show the termination of such abnormal veins.

Fig. 9.19

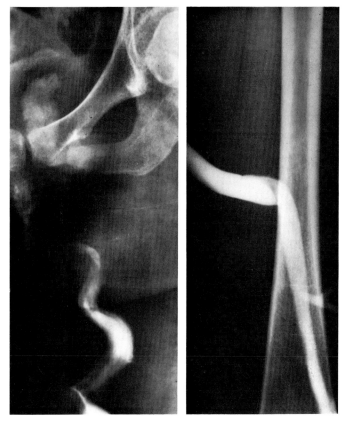

Fig. 9.20 Direct injection of a dilated valveless venous channel in Klippel-Trenaunay syndrome. The vein enters the perivesical plexus and this patient presented with haematuria.

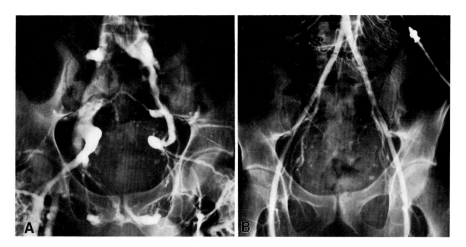

Fig. 9.21 Bilateral pertrochanteric intraosseous phlebogram in Klippel-Trenaunay syndrome (A) An abnormal venous channel enters the right common iliac vein. The channel is dilated at its termination. The internal iliac vein and the common iliac vein are also dilated on this side compared with the left. There are numerous phleboliths in the pelvis indicating that the full extent of the venous angioma has not been demonstrated. (B) Arteriogram in the same patient. The arteries are of normal size indicating that the venous dilatation is not due to arterio-venous fistula formation.

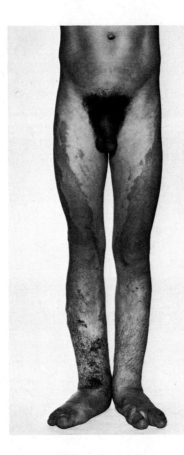

Fig. 9.22 A clinical photograph of a patient with Klippel-Trenaunay syndrome. The condition is usually unilateral but in this case is bilateral and in fact all four limbs can be affected. Note the naevus on both sides and the swelling of the soft tissues particularly of the lower legs. The right leg is more affected than the left, and on this side there is ulceration of the calf and ankle regions. A large varicose vein can be seen on the lateral aspect of the right thigh.

Klippel-Trenaunay syndrome

The primitive capillary plexuses of the flattened limb buds give rise in the embryo to peripheral border veins, one ventral and the other dorsal. During the second month the latter atrophies almost completely. In some patients the primitive dorsal or sciatic venous system persists in various forms usually as a large, lateral venous channel in the leg (Figs. 9.18, 9.19 and 9.20). A channel in this position

MALFORMATIONS 171

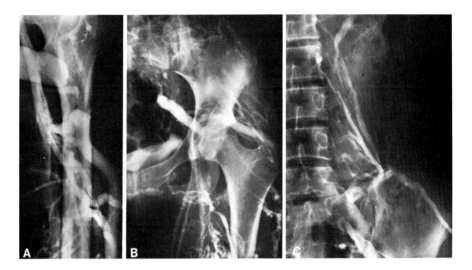

Fig. 9.23 Examples of abnormal veins in Klippel-Trenaunay syndrome. (A) The primitive lateral vein is joining the profunda femoris vein. (B) It joins the internal iliac vein (C) There are multiple connections including one to the inferior vena cava.

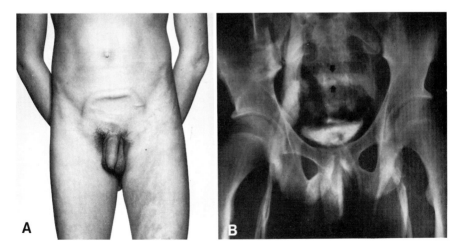

Fig. 9.24 The Klippel-Trenaunay syndrome. (A) A clinical photograph. Note the abnormal position and enlargement of a suprapubic vein. There is a naevus on the medial aspect of the left thigh. (B) Bilateral ascending phlebogram. The abnormal vein can be seen crossing the mid line (arrows). The flow within the vein was from left to right. The common iliac vein on the left side is absent. Ligation of such collateral veins is contraindicated.

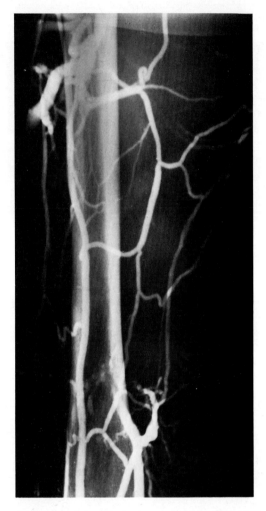

Fig. 9.25 Ascending phlebogram of the right leg in a patient with Klippel-Trenaunay syndrome. The popliteal and femoral veins are aplastic and the contrast medium passes through superficial veins.

together with other varicose veins is a constant feature of the syndrome described by Klippel and Trenaunay.[3] The other features of this syndrome include a naevus confined to one lower limb with hypertrophy of the bones and tissues of the affected limb. Occasionally both legs and the arms are affected. The original description made no mention of symptoms or signs of arterio-venous fistulae[3] (Fig. 9.21). Although there are variations in the extent and distribution of the

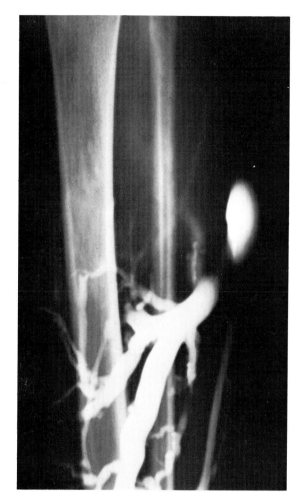

Fig. 9.26 Ascending phlebogram of the calf in a patient with Klippel-Trenaunay syndrome. The deep veins of the calf are atretic and the main venous flow is through an abnormal lateral venous channel. In this situation ligation of the superficial veins would increase leg oedema.

components of this syndrome it is generally easily recognised clinically (Fig. 9.22), and for this reason the eponym can be conveniently retained. Since the original description there have been a number of other reports of a similar syndrome[4,5,6,7,8,12,15,16] and it would appear to be the commonest venous angiodysplasia. While some authors have included arterio venous fistulae in the syndrome it is now generally accepted that the term Klippel-Trenaunay syndrome

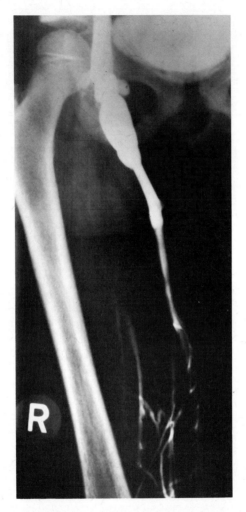

Fig. 9.27 Klippel-Trenaunay syndrome. A few hypoplastic veins can be seen in the lower part of the thigh. The main venous drainage is through the long saphenous vein which is dilated.

should not be used if there is an arterial element present.[6] Such a limitation in the definition is of practical value as the management and prognosis are different in arterio-venous malformations.

Phlebographically there is a lateral valveless dilated channel which may join the profunda femoris vein, the internal or external iliac veins, the common iliac vein or inferior vena cava, often by multiple connections (Fig. 9.23). This channel and other connecting varicose veins can be demonstrated by directly injecting the

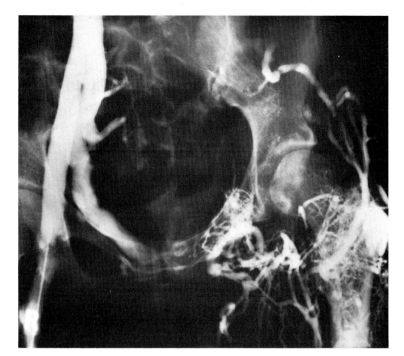

Fig. 9.28 Left pertrochanteric phlebogram and right femoral phlebogram in a patient with Klippel-Trenaunay syndrome. The upper part of the left femoral vein, the external iliac, the internal iliac vein and the common iliac vein are absent. The drainage of the leg is through a large pubic vein and also lateral veins bypassing the obstruction on the left side. The absence of these major deep veins cannot be artefactual because the intraosseous technique has been used.

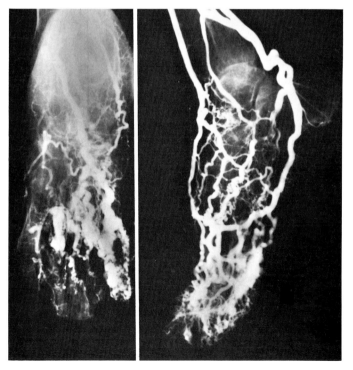

Fig. 9.29 An example of a diffuse angioma of the foot. A tight tourniquet has been applied above the site of the needle puncture in order to produce retrograde flow into the veins of the foot.

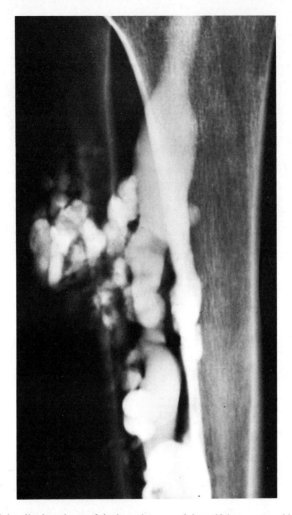

Fig. 9.30 A localised angioma of the lateral aspect of the calf demonstrated by ascending phlebography with a tourniquet above the knee to fill out the lesion.

dysplastic veins with the patient in the horizontal position and with the use of tourniquets to direct the contrast into the deep system if necessary.

The combination of the Klippel-Trenaunay syndrome with stenosed and occluded deep veins is said to be very rare.[14,15] However in the author's experience about 40 per cent of patients have aplasia or hypoplasia of some part of the venous system[6] (Figs. 9.24, 9.25, 9.26 and 9.27).

Technical factors may account for non-filling of the deep venous system in some patients because of difficulty in occluding the superficial system with a tourniquet in the presence of varicose veins and ankle swelling. Several of the examinations however were carried out using the intraosseous technique which

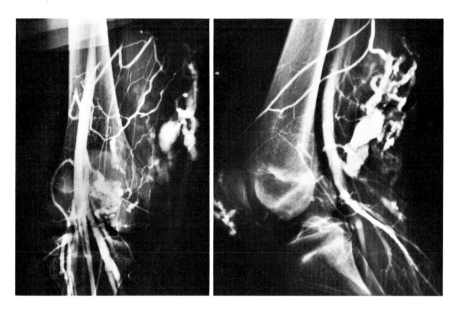

Fig. 9.31 A venous angioma of the lower thigh and the popliteal region demonstrated by direct injection with a tourniquet above the lesion.

always fills the deep venous system if it is patent (Fig. 9.28). It would therefore appear that abnormalities of the deep veins are rather more frequent than has been reported.

Short segment stenoses or occlusions can be treated by direct venous grafting and longer ones by saphenous vein bypass grafting of the Palma type.[11] In the presence of aplasia or hypoplasia of the deep venous system the conduit function of the superficial venous system must be preserved otherwise venous return will be further impaired resulting in oedema.

On the other hand if the deep venous system is shown to be intact incompetent communicating veins may be ligated together with large varicose venous channels. If soft tissue hypertrophy is a major feature of the syndrome, reducing operations for cosmetic and functional reasons may have to be considered.[1,17]

Other venous angiodysplasias

There are a very large variety of venous anomalies and malformations which may be localised or widespread (Figs. 9.29, 9.30, 9.31 and 9.32). Their extent and

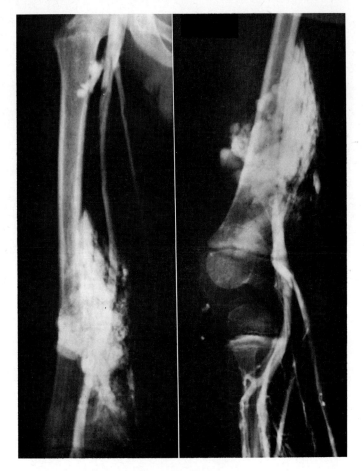

Fig. 9.32 A diffuse venous angioma of the thigh and knee region. The angioma extends into the knee joint and the patient presented with a haemarthrosis.

connections can be established by a combination of clinical examination and phlebography. Localised venous angiomas can often be removed surgically with good cosmetic results whereas the extensive plexiform types may have to be treated conservatively, or by sclerotherapy or therapeutic embolisation.

Readers who wish to study this complex subject in depth are referred to the detailed monograph by Schobinger.[13]

REFERENCES

1. Dodd H, Cockett F B 1976 The pathology and surgery of the veins of the lower limb. Churchill Livingstone, Edinburgh
2. Kinmonth J B, Young A E, O'Donnell T F Jr, Edwards J M, Lea Thomas M 1976 Mixed vascular deformities of the lower limbs. Br J Surg 63: 899

3. Klippel M, Trenaunay P 1900 Du naevus variquex osteohypertrophique. Arch Gen Med 185: 641
4. Langeron P, Vercauteren B 1970 Le syndrome de Klippel Trenaunay historique – Nosologie, aspects cliniques et pathogeniques. J Sci Med Lille 88: 561
5. Lea Thomas M, Andress M R 1971 Angiographic changes in venous dysplasias of the limbs. Am J Roentgenol 113: 724
6. Lea Thomas M, Macfie G B 1974 Phlebography in the Klippel Trenaunay syndrome. Acta Radiol 15: 43
7. Lindenauer S M 1965 Klippel Trenaunay syndrome. Ann Surg 162: 303
8. Lindenauer S M 1971 Congenital arteriovenous fistulae and the Klippel-Trenaunay syndrome. Ann Surg 174: 248
9. Malan E, Puglionisi A 1964 Congenital angiodysplasias of the extremities. J Cardiovasc Surg 6: 255
10. Negus D 1976 In: (ed) Dodd H, Cockett F B The pathology and surgery of the veins of the lower limbs. Churchill Livingstone, Edinburgh
11. Palma E C, Esperon R 1960 Vein transplants and grafts in the surgical treatment of the post phlebitic syndrome. J Cardiovasc Surg 1: 94
12. Parkes Weber F 1907 Angioma formation in connection with hypertrophy of limbs and hemi-hypertrophy. Brit J Derm 19: 231
13. Schobinger R A 1977 Periphere andysplasien. Verlag Hans Huber, Bern
14. Servelle M 1949 Les malformations congenitales des veines. Rev Chir 68: 88
15. Servelle M 1964 Syndrome de Klippel Trenaunay. Presse Med 72: 3323
16. Vollmar J 1974 Zur geschichte und terminologie der syndrome nach F P Weber und Klippel Trenaunay. Vasa 3: 231
17. Vollmar J 1979 In: (ed) May R Surgery of the veins of the leg and pelvis. Georg Theme Publishers, Stuttgart

10

Extrinsic compression

INTRODUCTION

Veins being thin walled are readily deformed, displaced or obstructed by adjacent structures. Such deformity may be symmetrical or asymmetrical, localised or diffuse, single or multiple, and smooth or irregular. The appearances are however generally non-specific and it is for this reason that phlebography has ceased to be the primary investigation of possible mass lesions.

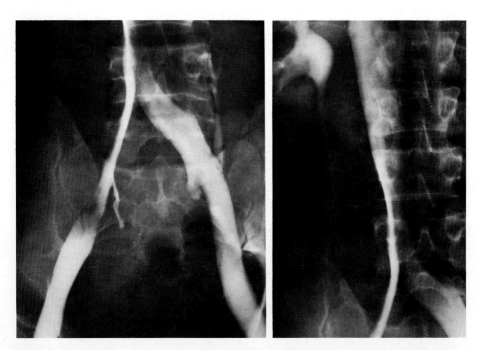

Fig. 10.1 Bilateral iliocaval phlebogram. This examination was carried out because the patient was complaining of recent onset of swelling of the right leg which was considered to be due to deep vein thrombosis. The phlebogram shows narrowing of the right external iliac and common iliac veins and the lower inferior vena cava having the appearance of external compression. Exploration and biopsy revealed that the cause of the external compression was extensive lymphoma. This phlebogram was the first investigation to indicate the cause of the patient's symptoms.

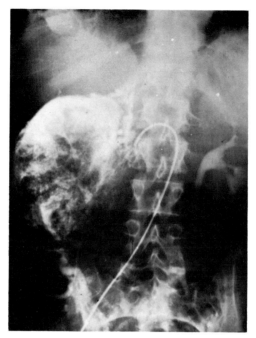

Fig. 10.2 A large volume selective renal arteriophlebogram. The right renal vein has not been demonstrated in this late phase and was assumed to be occluded by tumour thrombus. This was confirmed at operation. The large renal tumour of the lower part of the right kidney is shown by the pathological circulation.

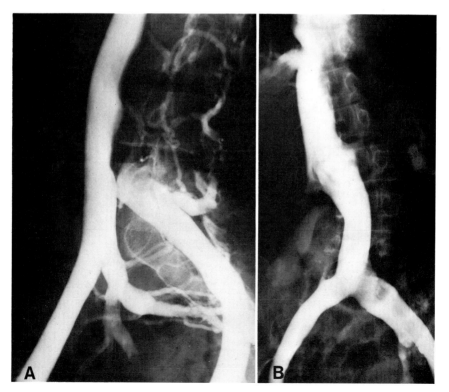

Fig. 10.3 Examples of normal structures deforming adjacent veins (A) Normal intervertebral discs indenting the inferior vena cava shown in the oblique projection. (B) A right pelvic kidney compressing the lower inferior vena cava.

182   PHLEBOGRAPHY OF THE LOWER LIMB

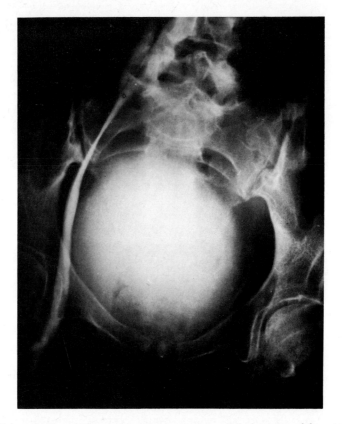

Fig. 10.4 Right ascending phlebogram showing narrowing and displacement of the external iliac and common iliac veins by an enlarged bladder. Such compression may lead to thrombosis and to pulmonary embolism in patients with chronic retention.

Phlebography undertaken in the investigation of swollen limbs to exclude venous thrombosis may however give the first indication of external compression and will provide a guide for further investigations (Fig. 10.1).

It is because of the limited use of phlebography in extrinsic deformities that this chapter is brief and concerned mainly with the interpretation of deformities commonly seen on phlebograms and with situations where the examination may be of help in diagnosis and management.

EXTRINSIC COMPRESSION 183

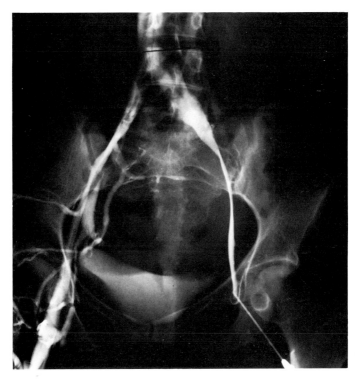

Fig. 10.5 The external iliac veins are displaced laterally and narrowed particularly on the left side by a pelvic mass which prevents filling of the left internal iliac vein. The mass is a large ovarian cyst.

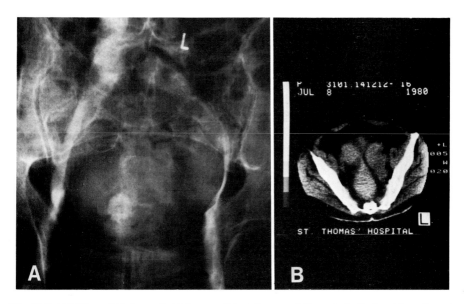

Fig. 10.6 (A) Bilateral iliofemoral phlebogram. The left external iliac vein is displaced laterally and narrowed by an extrinsic mass. (B) Computed tomogram of the pelvis showing the mass in the pelvis together with enlarged lymph glands. This was due to prostatic carcinoma.

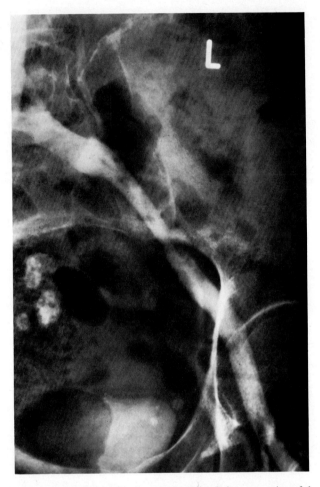

Fig. 10.7 Radiotherapy to left external iliac lymph glands has led to narrowing of the external iliac vein and leg swelling.

PHELBOGRAPHIC TECHNIQUES

The method of phlebography is determined by the suspected site of the lesion and any of the techniques described in Chapter 3 may be required although the most commonly used techniques are ascending and iliocaval phlebography.

An additional technique not previously mentioned, is large volume renal arteriophlebography which is necessary to demonstrate extension of renal tumour

EXTRINSIC COMPRESSION 185

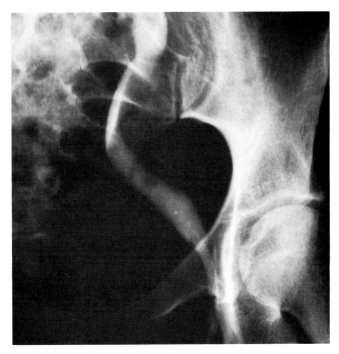

Fig. 10.8 Medial displacement of the external iliac vein by a lymphocoele.

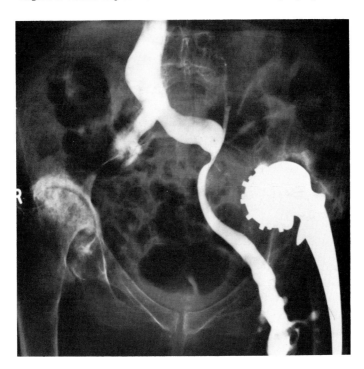

Fig. 10.9 This hip prosthesis has become centrally dislocated causing compression of the left external iliac vein. The narrowing of the vein with consequent stasis has led to deep vein thrombosis and swelling of the leg. A little recent thrombus can be seen in the femoral vein.

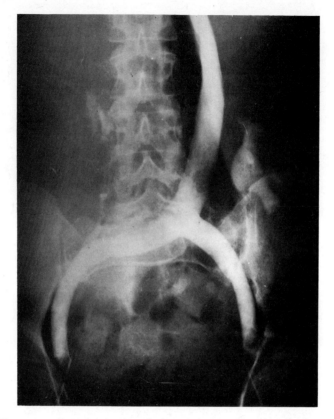

Fig. 10.10 There is considerable displacement of the inferior vena cava to the left by a large abdominal mass. This appearance is non specific but the mass was due to a right hypernephroma.

thrombus along the renal vein. The technique follows conventional selective renal arteriography when this confirms the presence of a malignant neoplasm. A second series of films is taken following a much larger quantity of contrast medium e.g. 50ml of meglumine iothalamate '280' with delayed films to show the venous phase (Fig. 10.2). Confirmation of tumour thrombus affects both the surgical approach and the patient's prognosis.[6]

ANATOMICAL CAUSES OF DEFORMITY

Arteries, bones and soft tissues are related to the venous system throughout its length and these structures may deform the venous system at any site (Fig. 10.3). Reference to compression by nearby or dilated arteries has already been made in Chapter 5.

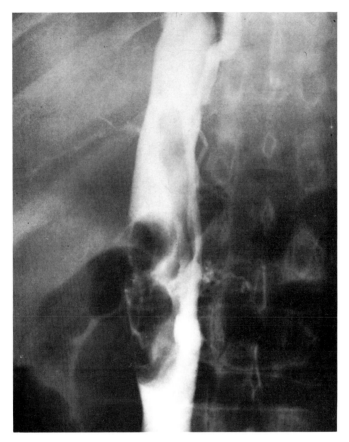

Fig. 10.11 This inferior vena cavogram shows tumour thrombus from a right renal carcinoma which has occluded the renal vein and extends into the inferior vena cava.

In pregnancy the uterus compresses the inferior vena cava.[7] As pregnancy progresses the lumen of the inferior vena cava may be obstructed in the supine position. This obstruction is not necessarily of significance but may be a cause of swelling of the legs during pregnancy.

PATHOLOGICAL CAUSES OF DEFORMITY

There are a vast number of pathological states causing venous deformity, the majority of which are found in the pelvis and abdomen.

Pelvic lesions

One of the commonest masses to displace and narrow the pelvic veins is the distended bladder in patients with outlet obstruction (Fig. 10.4). If the obstruction

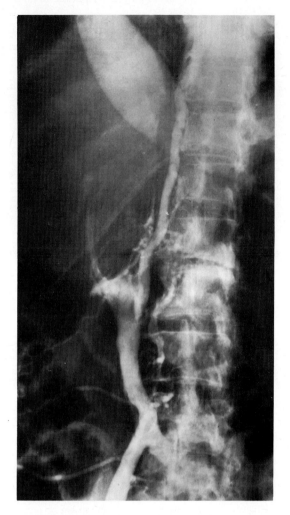

Fig. 10.12 The inferior vena cava is totally occluded by extension of a right hypernephroma. The venous drainage is now through the azygos system.

is not relieved the compression may lead to venous thrombosis and even pulmonary embolism.

In the female patient benign or malignant lesions of the pelvic viscera such as ovarian cysts or tumours of the uterus commonly involve the veins (Fig. 10.5). Likewise in the male pelvis carcinoma of the prostate is particularly liable to invade the iliac vein (Fig. 10.6). Carcinoma of the bladder or pelvic colon may also involve the pelvic veins in either sex.

Radiotherapy is frequently used to treat malignant lesions in the pelvis and the resulting fibrosis may cause venous stenosis or obstruction (Fig. 10.7).

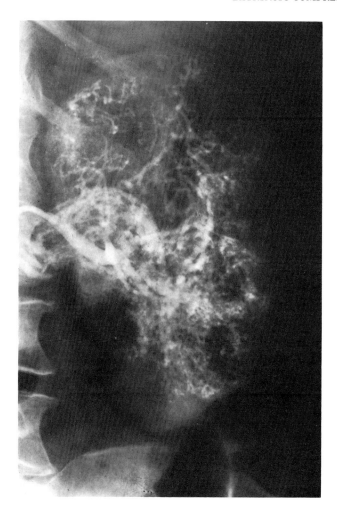

Fig. 10.13 Arteriophlebogram of the left kidney. The selective catheter can be seen in the renal artery. In this late phase the renal vein can be seen to be full of tumour thrombus demonstrated by the pathological circulation. The hypernephroma involves virtually the whole of the kidney.

Any other mass in the pelvis is likely to cause deformity of adjacent veins (Fig. 10.8 and 10.9).

Abdominal lesions

Any intraperitoneal viscus may affect the abdominal veins if sufficiently enlarged. For example an enlarged liver can elongate and narrow the hepatic portion of the inferior vena cava.

However the vast majority of lesions which involve the abdominal veins arise from retroperitoneal structures.

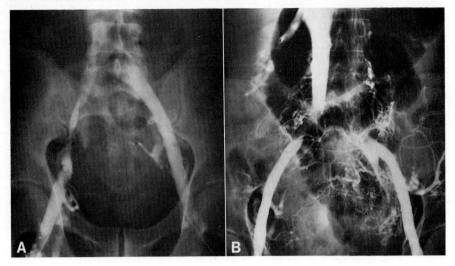

Fig. 10.14 Examples of retroperitoneal fibrosis. (A) There are stenoses of the right external and common iliac veins. (B) In this case the fibrosis almost occludes the inferior vena caval bifurcation.

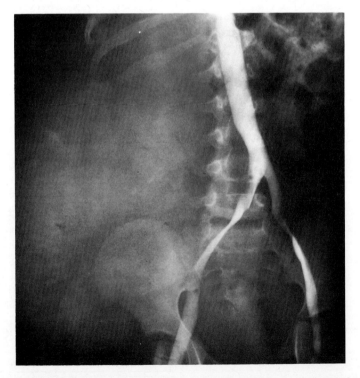

Fig. 10.15 An example of a very large right sided retroperitoneal tumour displacing the inferior vena cava and narrowing and displacing the right common iliac vein.

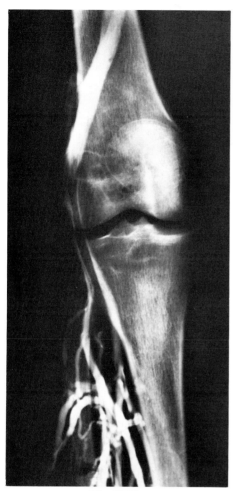 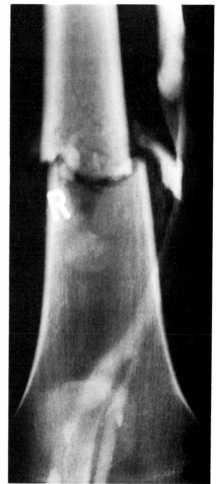

Fig. 10.16 **Fig. 10.17**

Fig. 10.16 A large popliteal cyst narrowing and displacing laterally the popliteal vein.

Fig. 10.17 A recent fracture of the lower part of the femur displacing and narrowing the popliteal vein and the lower part of the superficial femoral vein. The vein has been damaged by the trauma and a little contrast medium can be seen extravasating from the vein into the soft tissues.

The kidney

Renal masses compress, displace or invade the inferior vena cava.[9] Such caval involvement occurs most often in hypernephroma in the adult which may also involve the renal veins with tumour thrombus (Fig. 10.10, 10.11 and 10.12). It has already been pointed out that renal arteriophlebography may be useful in demonstrating the renal veins (Fig 10.13) but this may need to be supplemented by iliocaval phlebography. Similar changes may be seen in children with Wilm's tumour.[3]

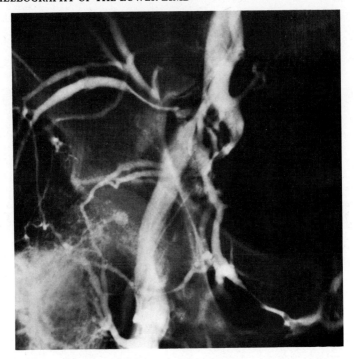

Fig. 10.18 The common femoral vein can be seen on this intraosseous phlebogram to be indented from the medial side by a mass of enlarged lymph nodes.

The adrenal
Tumours of the adrenal gland are particularly important in children[1,3,8] but adrenal carcinoma also occurs in adults. Masses confined to the adrenal glands themselves including adrenal hypertrophy are unlikely to show on inferior vena cavography.

The pancreas
Distortion of the upper inferior vena cava has been observed with both pseudocysts and carcinoma of the pancreas.[1,5]

Arterial
Tortuosity, arteriomegaly and aneurysms are frequent causes of indentations and displacement of the inferior vena cava because of their close anatomical relationship.

Retroperitoneal fibrosis and other abdominal lesions
Retroperitoneal fibrosis most frequently occurs around the pelvic brim and for this reason the condition may involve the iliac veins either locally or diffusely (Fig. 10.14). Similar irregularity and narrowing of the inferior vena cava may be

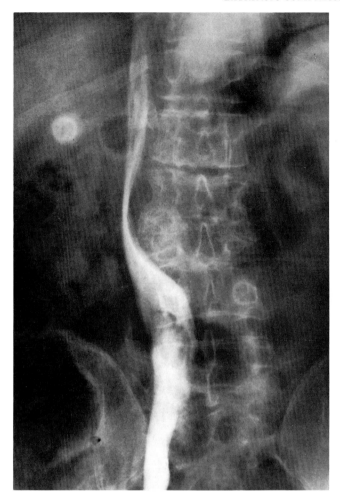

Fig. 10.19. A right iliocavogram. The inferior vena cava is narrowed and indented on its medial side by an old spinal fracture.

seen with inflammatory aneurysms but in these there is always additional displacement of the cava. Computed tomography can be useful in distinguishing the two conditions especially when the aortogram appears relatively normal.[10]

Retroperitoneal tumours may displace, deform or actually invade the inferior vena cava, however the appearances are often not specific (Fig. 10.15).

Lesions causing deformity at other sites

Whilst pelvic and abdominal lesions account for the majority of extrinsic abnormalities shown by phlebography lesions elsewhere may deform the phlebogram. Such deformity may occur in the leg, groin or parapelvic region (Figs. 10.16, 10.17, 10.18, 10.19).

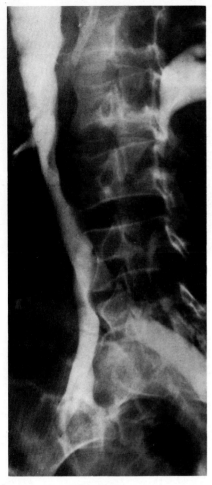 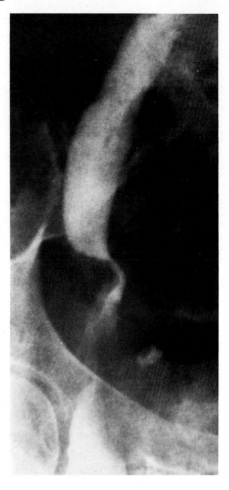

Fig. 10.20   **Fig. 10.21**

Fig. 10.20 Multiple lobulated indentations of the inferior vena cava typical of enlarged lymph nodes. The patient had Hodgkin's disease.

Fig. 10.21 There is an irregular indentation on the lateral aspect of the external iliac vein due to an adjacent malignant lymph node.

LYMPH NODES

Enlarged lymph nodes present a characteristic lobulated appearance of the inferior vena cava and iliac veins.

Glands involved by lymphoma tend to produce widespread indentations along the iliocaval systems (Fig. 10.20) whereas metastatically involved glands are usually more localised (Fig. 10.21). Common primary sites for such metastatic nodes include the ovary, uterus and testis.

The information obtainable by iliocaval phlebography about the state of the

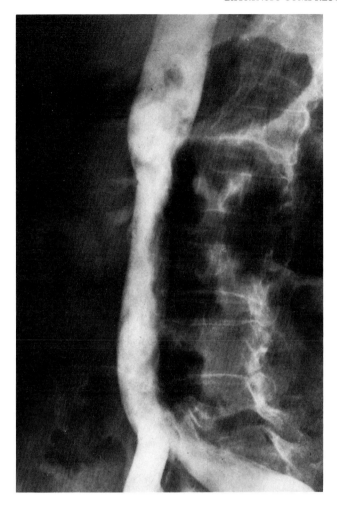

Fig. 10.22 An ascending iliocavogram. There are small irregular defects on the postero-medial aspect of the inferior vena cava. This represents malignant involvement of the inferior vena cava by lymph nodes. This abnormality only showed in the oblique projection.

lymph nodes is limited because significant enlargement is necessary before deformity is produced. The iliocavogram may need to include straight, lateral and varying degrees of obliquity to show deforming lesions (Fig. 10.22). Another disadvantage is that the left para-aortic glands lie at a distance from the inferior vena cava and considerable enlargement may occur without affecting the cavogram.

THE VALUE OF PHLEBOGRAPHY

As all lesions which compress veins predispose to stasis secondary venous thrombosis may be the only demonstrable abnormality (Fig. 10.23). A further

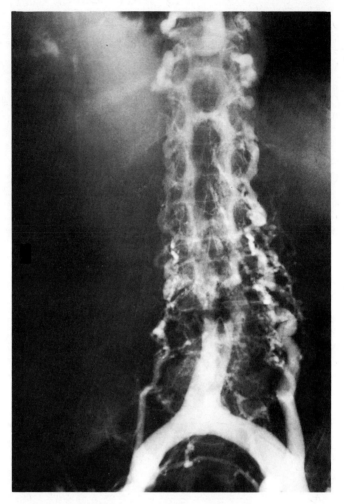

Fig. 10.23 This patient had a retroperitoneal sarcoma causing thrombotic occlusion of the lower inferior vena cava. The sharp cut-off is not typical of primary venous thrombosis and raises the possibility of an underlying malignancy resulting in secondary thrombosis of the inferior vena cava.

limitation of phlebography is that it is frequently very difficult to distinguish between benign and malignant causes of venous deformity.

Compression due to a benign lesion commonly gives a smooth defect whereas that due to malignancy is often irregular in outline due to invasion of the vein wall.

In order to obtain an accurate diagnosis other radiological investigations such as arteriography (Fig. 10.24), lymphangiography, (Fig. 10.25) and computed tomography (see Fig. 10.6) are frequently necessary.

EXTRINSIC COMPRESSION 197

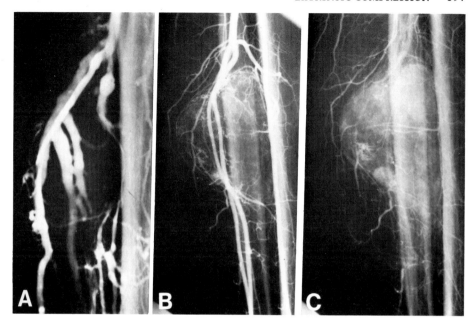

Fig. 10.24 (A) Ascending phlebogram of the calf. The posterior tibial veins are displaced anteriorly and the short saphenous vein displaced posteriorly indicating a mass in the calf. However, the phlebogram gives no indication of the nature of the mass. (B) Femoral arteriogram (arterial phase). A little pathological circulation can be seen within the mass. (C) (Late phase). In this film the mass is clearly outlined and the extensive malignant circulation throughout it is confirmed. This lesion was shown at biopsy to be a fibrosarcoma arising from the intraosseous membrane.

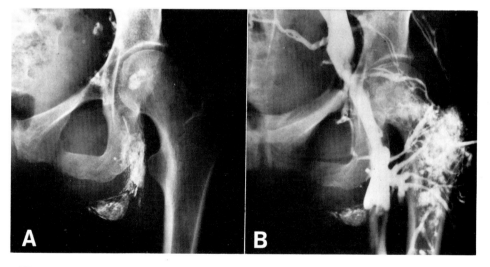

Fig. 10.25 This patient presented with a mass in the left groin following a stripping operation of the long saphenous vein for varicose veins. (A) Lymphadenogram. This shows displacement of the lymph nodes in the groin around the soft tissue mass. A small fleck of lipiodol is present within the mass strongly suggesting that this is a lymphocoele. The diagnosis was confirmed at operation. (B) Intraosseous phlebogram. This examination gives no further information as to the nature of the mass. There is very slight bowing of the common femoral vein around the mass.

Because phlebography is easy to perform and carries a low morbidity it is a worthwhile procedure in patients with masses or unexplained swelling of the legs.

For a more detailed description of the causes of extrinsic deformities of the vena cava and tributaries readers are referred to the monographs by Ferris et al,[4] and Chermet and Bigot.[2]

REFERENCES

1. Allen J E, Morse T S, Frye T R, Clatworthy Jr H W 1964 Vena cavograms in infants and children. Ann Surg 160: 568
2. Chermet J, Bigot J H 1980 Venography of the inferior vena cava and its branches. Springer-Verlag, Berlin
3. Ducharme J C, Ellis F 1964 Inferior vena cavogram; an aid in the diagnosis of abdominal tumours in children. J Assoc Clin Radiologists 15:38
4. Ferris E J, Hipona F A, Kahn P C, Phlipps E, Shapiro J H 1969 Venography of the inferior vena cava and its branches. The Williams and Wilkins Co, Baltimore
5. Filler R M, Harris S H, Edwards E A 1962 Characteristics of the inferior-cava venogram in retroperitoneal cancer. New Engl J Med 266: 1194
6. Lea Thomas M, Lamb G H R 1979 The value of large volume selective arteriophlebography of the renal veins in the preoperative assessment of renal carcinoma. Brit J Urol 51: 78
7. Samuel E 1964 The inferior-vena cavogram in pregnancy; radiologic aspects. Proc Roy Soc Med 57: 702
8. Tucker A S 1965 The roentgen diagnosis of abdominal masses in children; intravenous venogram vs. inferior vena cavography. Am J Roentgenol 95: 76
9. Wright F W 1965 Cavography in the assessment of renal tumours. Brit J Urol 37: 380
10. Young A E, Lea Thomas M, Wright C H 1980 Assessment of abdominal aortic aneurysms by computed tomography. Brit Med J 280: 765

11

Miscellaneous uses of phlebography

This chapter discusses a number of venous conditions and variations of phlebographic technique which do not readily form part of the main subdivisions of this monograph. The subjects are unrelated to each other but it is felt that they should be briefly considered for the sake of completeness.

OPERATIVE PHLEBOGRAPHY

During thrombectomy
It is generally recognised that iliofemoral thrombosis is a common and serious event both as regards the possibility of fatal pulmonary embolism and the development of the post thrombotic state. These considerations are discussed in more detail in Chapters 6 and 7.

The aims of venous thrombectomy in this situation are to restore patency of the iliofemoral segment so that symptoms of acute and chronic venous insufficiency may be minimised, and to reduce the risk of embolism.[14]

In its simplest form peroperative phlebography involves the injection of contrast distal to the site of thrombectomy and either watching the flow of the medium through the thrombectomised segment with a portable image intensifier, or taking a film as the contrast is injected.

Mavor and Galloway[14] recommend dividing the superficial circumflex iliac vein and preserving the portion leading to the long saphenous vein so that a catheter may be passed through it to the external iliac vein to facilitate postoperative phlebography.

Other perioperative situations
One of the most useful situations for the use of perioperative phlebography occurs in bypass surgery. Not only is it necessary to demonstrate the venous anatomy beforehand but it is also desirable to confirm the functional success of the reconstructive procedure. Phlebography may also be required following various surgical procedures performed for varicose veins, particularly if surgical damage to the deep veins is suspected (Fig. 11.1) or if there is rapid and unexpected recurrence of varicosities. Venous angiodysplasias often require preoperative phlebography to demonstrate their extent and connections, as well as post excision phlebography to confirm complete removal (see Chapter 9).

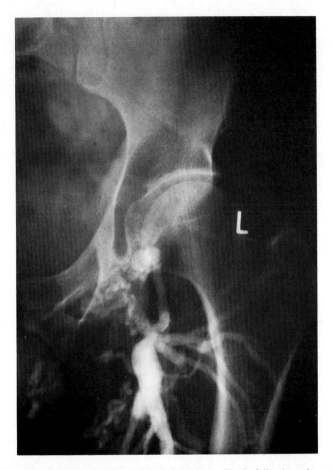

Fig. 11.1 This patient developed swelling of the left leg immediately following a long saphenous ligation and stripping operation for varicose veins. The phlebogram reveals that the cause of the swelling of the leg is due to inadvertent ligation of the femoral vein.

Technique

Since elaborate serial changers and sophisticated television monitoring systems are not often available in operating theatres, single film radiography is usually all that can be carried out. For this reason the timing of the film is critical. A relatively large volume of contrast medium, i.e. 50ml of 60% meglumine iothalamate, should be injected continuously by hand and a film taken towards the end of the injection. In this way diagnostic results can be obtained with the minimum of delay, avoiding repeated injections if the critical phase is missed.

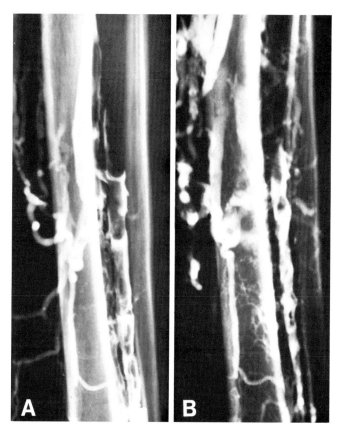

Fig. 11.2 (A) This ascending phlebogram shows recent venous thrombus in the peroneal veins. The proximal part of the lateral peroneal vein is totally occluded by thrombus. At this stage heparin therapy was started. (B) A repeat phlebogram after 7 days of heparin shows that there has been no propagation of the thrombus and there has been some contraction. It is particularly important to prevent spread of thrombus from the calf veins into the popliteal and more proximal veins. It is destruction of the valves in the latter veins which give rise to the post thrombotic syndrome.

ANTICOAGULANT AND THROMBOLYTIC THERAPY

Heparin is usually given in the initial therapy of deep venous thrombosis while oral anticoagulation is being established. Heparin is not a thrombolytic agent but in the correct dosage prevents propagation of thrombus. This may stop established thrombus reaching lethal proportions. It will also minimise the extent of post thrombotic damage to the valves and the veins themselves, which is particularly important in the large veins above the calf. (Fig. 11.2).

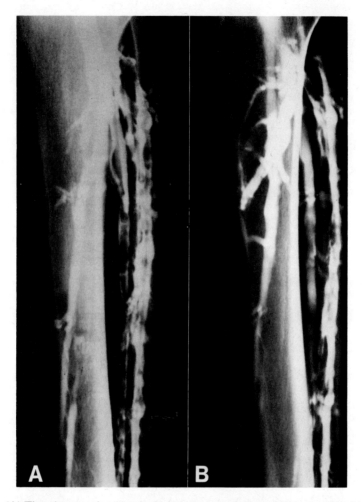

Fig. 11.3 (A) There is recent thrombus in the lateral peroneal vein and in the anterior tibial veins. (B) After streptokinase the peroneal vein has become normal and valves are visible. The thrombus in the anterior tibial veins is less.

Thrombolytic agents such as streptokinase and urokinase will lyse thrombus provided it is in a free flowing bloodstream.[2] The phlebogram is repeated on about the 5th day of a standard course of therapy. Further thrombolytic therapy is determined by comparison of the two examinations to assess progress (Figs. 11.3, 11.4 and 11.5).

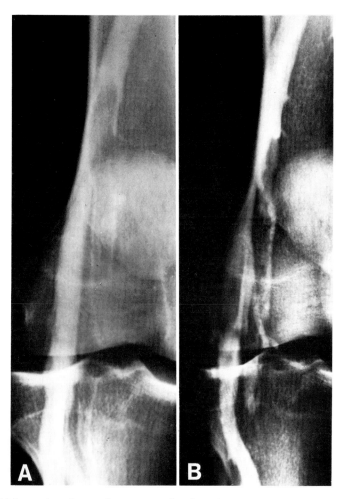

Fig. 11.4 (A) Loose thrombus can be seen extending from the short saphenous vein into and up the popliteal vein. (B) After streptokinase the thrombus in the short saphenous vein has considerably retracted and the popliteal vein above the entry of the short saphenous vein is now patent. The thrombus previously present in the popliteal vein has been lysed.

CLINICAL TRIALS

Phlebography has a very small morbidity and it is therefore justifiable to use it for clinical trials with the patient's understanding and permission. Thus it has been employed in assessing the radiological progression of deep vein thrombosis[10] and in comparing the results of surgical and medical treatment of venous thrombosis.[19] Phlebography has been used to assess the generalised and local side effects of the

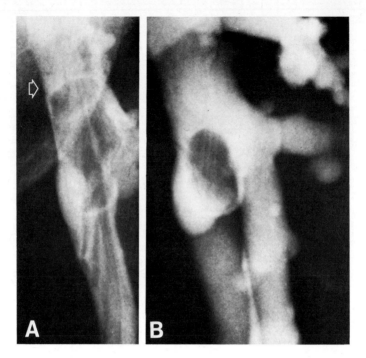

Fig. 11.5 (A) There is loose thrombus in the common femoral vein surrounded by contrast medium indicating that it lies freely in the blood stream. The sharp cut-off of the thrombus indicates that an embolus has already occurred. (B) After streptokinase there has been partial lysis of the thrombus and a valve cusp is now clearly seen below it.

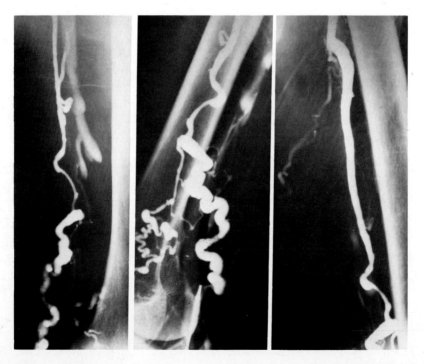

Fig. 11.6 Varicography. A direct injection of contrast medium into a varicose vein has been carried out to show its extent.

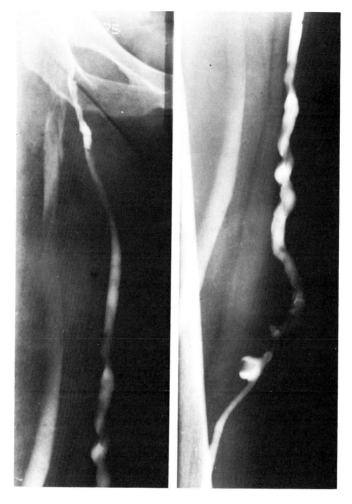

Fig. 11.7 Superficial phlebogram. The long saphenous vein has been injected with contrast to show its patency. The vein is slightly varicose but acceptable for use in bypass surgery.

new low osmolality contrast media by comparing them with the conventional hyperosmolar media in current use.[1, 11, 18]

Despite the side effects, about half the patients asked to undergo repeat phlebography for clinical trials do so.

SUPERFICIAL PHLEBOGRAPHY

Reference has already been made to the use of superficial phlebography in Chapters 3, 8, and 9. The essence of the technique is to inject directly the long or

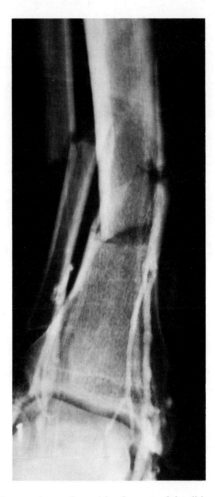

Fig. 11.8 Ascending phlebogram in a patient with a fracture of the tibia and fibula. The examination was carried out to show any damage to the venous system. The posterior tibial vein is narrowed at the fracture site but there is no evidence of extravasation of the contrast medium.

short saphenous vein or one of their tributaries at or around the ankle with the patient in the supine position and without tourniquets to direct the contrast into the deep venous system. As the contrast medium passes centrally its progress is followed on a television monitor and spot films taken of the superficial veins. Apart from the use of superficial phlebography to demonstrate varicose veins (Fig. 11.6) and to demonstrate venous angiomas and their tributaries together with any deep connections, there are other situations in which superficial phlebography may be useful. Superficial phlebography may be required to

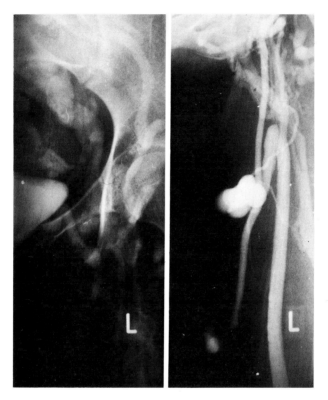

Fig. 11.9 The post thrombotic syndrome following penetrating trauma to the left thigh. The common femoral and profunda femoris veins together with the iliac veins have irregular lumens and collateral circulation is present. The superficial femoral vein appears normal. A venous aneurysm is demonstrated which is also the result of the trauma.

demonstrate a suitable vein for an arteriovenous shunt in patients with chronic renal failure.[13] Another use is to identify suitable segments of veins for use in arterial bypass surgery both in peripheral arterial disease and in coronary artery disease (Fig. 11.7).

VENOUS TRAUMA

Trauma to the venous system is common but produces less dramatic and immediate effects than arterial trauma. For this reason it has received less

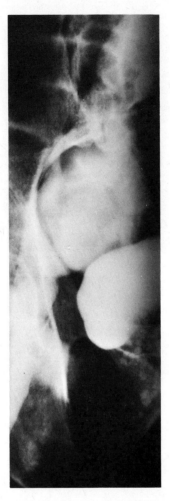

Fig. 11.10 Right iliac phlebogram showing a large aneurysm of the external iliac vein as a result of a fractured pelvis.

consideration, management usually being confined to tying off bleeding veins. Improved phlebographic techniques[8, 16] have drawn attention to the late sequelae of the ligation of main veins especially in the form of the post thrombotic syndrome, and more recently interest has been focussed on reconstructive procedures.[17]

Phlebography may be used to assess the site and the extent of an acutely injured vein. It can also be used at a later stage following venous injury to assess the long term effects of emergency therapy such as major vein ligation. Venous injuries may result from penetrating trauma including surgery, blunt trauma, compression by haematomas or fractures (Fig. 11.8) and direct injury due to stretching

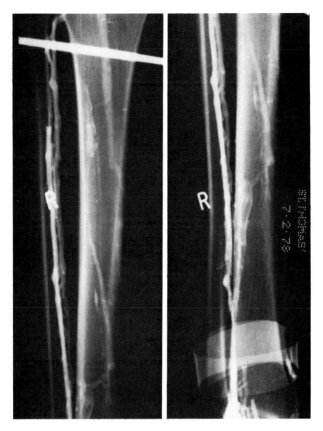

Fig. 11.11 A modification of phlebographic technique in trauma. This patient had a fracture of the femur and was on traction. The ascending phlebogram was carried out with the patient supine with no table tilt and the contrast directed into the deep venous system by a tight ankle tourniquet. The phlebogram shows recent thrombus in the calf veins.

associated with dislocations. The long term effects which concern the vascular surgeon are the post thrombotic syndrome (Fig. 11.9), arteriovenous fistulae, and aneurysms (Fig. 11.10). In these acute and chronic situations phlebography plays an important part in management.[15]

The technique of phlebography has to be tailored to the particular circumstance (Fig. 11.11). Injured patients are usually unable to stand so that steep table tilts to fill the deep venous system cannot be employed. In this case tourniquets around the ankles or proximal to the site of injection are used to direct the contrast into the deep venous system. A single film after the injection of 50 to 100ml of contrast medium is often all that can be obtained in the casualty department or operating theatre.

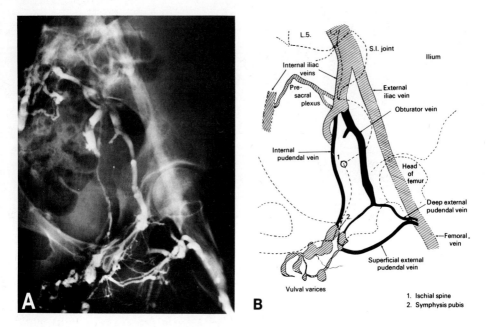

Fig. 11.12 (A) Vulval phlebogram by direct injection of a varix. (B) Schematic drawing of this phlebogram showing the main anatomical connections of the varices.

Iliofemoral thrombosis is common after hip replacement surgery, but in addition postoperative immobility results in venous stasis with a possibility of calf vein thrombosis. For this reason an ascending phlebogram in suspected venous thrombosis in these patients should be carried out to show the whole of the deep venous system of the legs and pelvis.

VULVAL AND PELVIC VARICES

Vulval varices

Varicose veins of the vulva occur in about 2 per cent of pregnant women and in 20 per cent of these the varicosities remain after delivery and tend to progress with subsequent pregnancies.[6] The symptoms which are worse at the time of menstruation consist of aching, discomfort and swelling of the vulva and thighs, increasing as the day goes on but eased by lying down. Sometimes the symptoms are sufficiently severe to require surgical treatment either during or after pregnancy. In some instances the varicose veins of the vulva are so severe that a normal vaginal delivery would be extremely hazardous and Caesarian section has to be carried out.

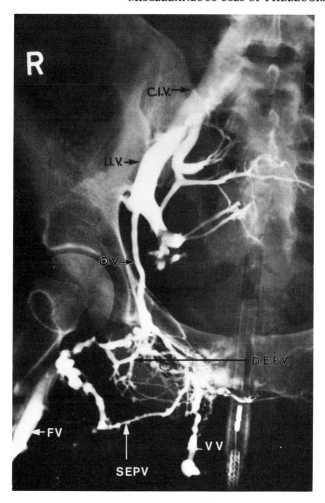

Fig. 11.13 Vulval varices (VV) and their connections in a straight projection. (CIV = common iliac vein; IIV = internal iliac vein; OV = ovarian vein; DEPV = deep external pudendal vein; SEPV = superficial external pudendal vein; FV = femoral vein). A radiation dose meter is in the vagina.

As the internal iliac veins are the main drainage veins of the uterus the increased pressure flow from the organ, when gravid, is transmitted to the veins of the vulva and the inner side of the thigh which consequently dilate and become varicose.

As the surgical treatment of vulval varices consists of freeing and excision of the main visible veins and their tributaries[5] a knowledge of their anatomy is helpful. The varicose veins drain predominantly into the internal pudendal and obturator veins which are tributaries of the internal iliac vein (Fig. 11.12). Sometimes there is also a contribution from the superficial pudendal branch of the saphenous vein (Fig. 11.13).

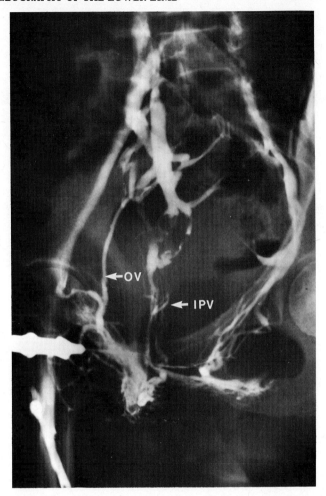

Fig. 11.14 Intraosseous phlebogram into the right os pubis. The 40° oblique projection used in this case clearly separates the obturator vein (O.V) from the internal pudendal vein (I.P.V). The internal pudendal vein shows recanalisation changes resulting from thrombosis. This has probably contributed to the development of the vulval varices. As can be seen it is often possible to demonstrate the contra-lateral side from a single injection.

Phlebography of vulval varices can be carried out by direct venepuncture of a vulval varix through a 21 gauge 'butterfly' needle which then is strapped into position[4] but occasionally a cut down onto a vein draining the varices is required.[12]

A test injection of contrast medium such as meglumine iothalamate '280' is made under television control to confirm the correct position of the needle and to ensure there is no extravasation. The patient is placed in a 40° oblique position being turned towards the side under examination (Fig. 11.14). A serial changer is

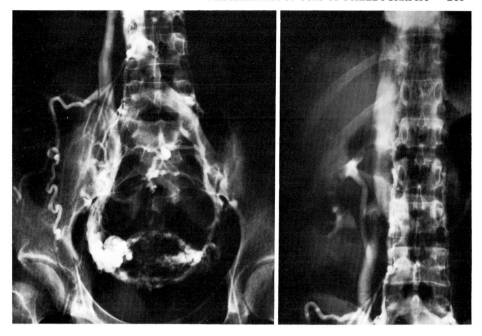

Fig. 11.15 Primary pelvic varicosities. This iliocaval phlebogram shows extensive periuterine dilated veins and the right ovarian vein is considerably enlarged.

positioned under the patient's pelvis and the tube centred two inches above the pubic symphysis so that the field includes the greater trochanter and the iliac crest. About 50ml of contrast medium is injected continuously by hand and four films exposed during the injection and a fifth 10 seconds after its completion.

Intraosseous pubic phlebography may be required if no suitable venepuncture site is present. A sternal puncture needle is introduced into the medullary cavity of the body of the pubis and a test injection of contrast is made under television control to check the position of the needle. Twenty ml of contrast is injected by hand and three to four films are taken at 2 second intervals. The positioning of the patient and the centreing are the same as described for direct phlebography (see Fig. 11.14).

Female pelvic varices

Varicosities of the uterovaginal plexus and of the broad ligament are very common. (Fig. 11.15). They may be the basis of a series of vague complaints including a sensation of heaviness in the pelvis, dysmenorrhea, symptoms of vulval, thigh and leg varices, and swelling and discomfort in the limb commonly referred to as the pelvic congestion syndrome.[14]

Pelvic varicosities are frequently associated with vulval varices.[14] In some patients they may result from extensive pelvic visceral collateral circulation which can develop from iliac vein obstruction (Fig. 11.16). The method of phlebography

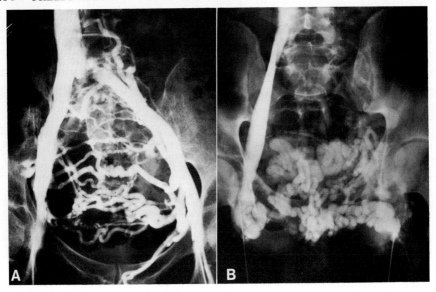

Fig. 11.16 Examples of secondary pelvic varicosities. (A) There are moderately extensive pelvic visceral collateral veins probably associated with a left common iliac vein stenosis. (B) In this patient there are massive collaterals across the floor of the pelvis due to left external and common iliac vein occlusion. Operations such as hysterectomy will destroy these veins and should be avoided.

has been described by Helander and Lindblom[7] who injected contrast medium into catheters introduced via the femoral veins using the Seldinger technique and then advanced proximally to the union of the external and internal iliac veins. Simultaneous compression of the inferior vena cava was recommended. A catheter may also be advanced more proximally and a contrast injection made into the left renal vein. The ovarian veins and varicose plexus in the parametrium are thus filled retrogradely. This method has been considered to be a considerable improvement on transuterine phlebography.[3]

The author finds the simpler procedures of percutaneous femoral or pertrochanteric intraosseous injections, usually provide sufficient information for surgical or other management.

Good therapeutic results have been recorded following hysterectomy[4] but this may be contraindicated if the parametrial plexus is functioning as a collateral pathway in iliac vein obstruction.[5] Another surgical procedure which may be successful is ligation of the left ovarian vein.

VARICOSE OVARIAN VEINS

Varicose ovarian veins may develop as collaterals in iliofemoral vein thrombosis. The vein may enlarge to such an extent that distal thrombus may pass through it

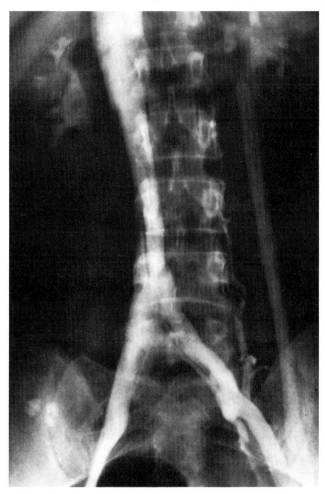

Fig. 11.17 Enlargement of the ovarian vein in iliocaval obstruction. In this case the narrowing of the inferior vena cava and the left common iliac vein is due to past thrombosis. The greatly enlarged left ovarian vein is of sufficient size to allow passage of an embolus. Such collaterals are more likely to occur after therapeutic *ligation* than *plication* of the inferior vena cava.

(Fig. 11.17). Such an enlargement also follows inferior vena caval ligation and less commonly plication. For this reason it is probably wise to ligate the vein prophylactically when carrying out caval interruptive procedures.

REFERENCES

1. Albrechtsson U, Olsson C G 1979 Thrombosis after phlebography; A comparison of two contrast media. Cardiovasc Radiol 2: 9
2. Browse N L, Lea Thomas M, Pim H P 1968 Streptokinase and deep vein thrombosis. Brit Med J 3: 717

3. Chidekel N, Edlundh K O 1968 Transuterine phlebography with particular reference to pelvic varicosities. Acta Radiol 7: 1
4. Craig O, Hobbs J T 1975 Vulval phlebography in the pelvic congestion syndrome. Clin Radiol 25: 517
5. Dodd H, Cockett F B 1976 The pathology and surgery of veins of the lower limb 2nd edn. Churchill Livingstone, Edinburgh
6. Dodd H, Payling-Wright H 1959 Vulval varices in pregnancy. Brit Med J 1: 831
7. Helander C G, Lindblom A 1959 Retrograde pelvic venography. Acta Radiol 51: 402
8. Kappert A, May R 1968 Das postthrombotische Zustandibild der Extremitäten. Huber, Bern.
9. Lea Thomas M, Fletcher E W L 1967 The venous drainage of the vulva. J Anat - Lond 101: 634
10. Lea Thomas M, McAllister V 1971 The radiological progression of deep vein thrombosis. Radiology 99: 37
11. Lea Thomas M, Walters H L 1979 Metrizamide in venography. Brit Med J 2: 1036
12. Lea Thomas M, Fletcher E W L, Andress M R, Cockett F B 1967 The venous connections of vulval varices. Clin Radiol 18: 313
13. Lea Thomas M, Rappaport A S, Wing A J 1974 Arteriography and phlebography of the legs in planning shunt and fistula operations in patients with chronic renal failure. Am J Roentgenol 121: 551
14. Mavor G, Galloway J M D 1967 The iliofemoral segment as a source of pulmonary emboli. Lancet 1: 871
15. Nylander G, Semb H 1972 Veins of the lower part of the leg after tibial fractures. Surg Gynec Obstet 134: 974
16. Shea P C Jr, Robertson R L 1951 Late sequelae of inferior vena cava ligation. Surg Gynec Obstet 93: 153
17. Vollmar J 1979 Venous trauma. In: May R (ed) Surgery of the veins of the leg and pelvis. Thieme Publishers, Stuttgart.
18. Walters H L, Clemenson J, Browse N L, Lea Thomas M 1980 ^{125}I fibrinogen uptake following phlebography of the leg. Radiology 135: 619
19. Young A E, Lea Thomas M, Browse N L 1974 Comparison between sequelae of surgical and medical treatment of venous thrombo-embolism. Brit Med J 4: 127

Index

Abdominal veins, deformities, causes, 189
Adrenal tumours, caval involvement, 192
Air bubbles, inadvertent injection producing filling defects, 75
Air embolism complicating phlebography, 63
Anatomy of venous system, 6–21
Aneurysm, venous, 165, 166
Angioma, venous, 153–157, 175, 178
Ankle, venous ulcer, 116
Anticoagulant therapy, phlebography during, 201
Anticoagulants and phlebography, 57
Artefacts, 66–84
  differential diagnosis, 82
  extrinsic pressure defects, 79
  iatrogenic, 81
Arteries, compression, 79
  deformities causing extrinsic venous compression, 192
Arteriophlebography, 153
Arteriovenous fistula, 157, 158

Blood
  nonopacified, causing artefact, 72
  pressure, venous, measurement, 132
Bolus technique to show iliac veins and inferior vena cava, 36
  risk of pulmonary embolism in, 64
Bone infarction after intraosseous phlebography, 62
  injection, in intraosseous phlebography, 42–48
Broad ligament, varicosities, 213
'Butterfly' needle, 27, 28
Bypass surgery, phlebography during, 199

Calf
  communicating veins, 17
  deep veins, post thrombotic state, 116
  muscle veins, 13
    artefacts, 68, 69, 70
  vein thrombosis, 33, 96
  phlebography in, 209
Cannula, plastic, 27
Cardiac complications of phlebography, 63
Clinical trials, phlebography in, 203

Complications, 54–65
  local, 56
  systemic and idiosyncratic, 55
Compression, extrinsic, 180–198; see Veins, compression, extrinsic
Contrast medium
  extravasation, local, 57–60, 62
  for ascending phlebography, 31
  hyperbaric, nature of, 1
  in demonstration of venous thrombosis, 1
  injection, 31
  layering defects, 69
  mixing defects, 69
  osmolality, reduction, 54
  pain due to, 56
  side effects, 54, 205
Cuff, pneumatic, 145
  supra-malleolar, inflated, 1

Embolism, air, complicating phlebography, 63
  fat, after intraosseous phlebography, 62
  pulmonary
    complicating phlebography, 64
    fatal, site of origin, 106
    incidence, 101
    mortality rate, 103, 105
    phlebography in, 103
    recurrence, prevention, 104
    source, 3
Examination in semi-erect position, 1
Exercise phlebography, 38
Extravasation, local, of contrast media, 57
Extrinsic arterial pressure defects, 79

Fat embolism after intraosseous phlebography, 62
Femoral veins, 23
  common, 13, 18
    artefacts, 73, 77, 78
  deep, 12
  ligation, accidental, 200
  superficial, 7, 10–13, 15, 18, 23
    artefacts, 71, 72
    occlusion, post thrombotic, 122, 123
    pressures, post thrombotic, 134
    thrombosis, 204

Fibrosis, retroperitoneal, involving iliac veins, 192
Fistula, arteriovenous, 157, 158
Foot
  veins
    phlebography, 36, 37
    pressures, post thrombotic, 134
    venous system, 8

Gastrocnemius veins, 12–14

Haemarthrosis complicating intraosseous phlebography, 61
Haematoma due to iliocaval phlebography, 63
Heparin, 57, 201
Historical aspects, 1–5
Hunter's canal incompetent communicating vein, 145, 148
Hyaluronidase injection after extravasation of contrast medium, 60
Hypertension, pulmonary, complications of phlebography in, 63

Iliac
  cross-over phlebography, 52
  veins
    aneurysms, 208
    artefacts, 70
    common, 7, 15, 20, 44
      artefacts, 77–79
      filling, 35
      stenosis, 214
    compression defect, 125, 126, 180, 182–185
    deep circumflex, 20
    external, 12, 18–20
    internal, 20, 30
    obstruction, collaterals, 126, 128
      reconstructive surgery in, 134, 135
    perforation during iliocaval phlebography, 63
    phlebography, bolus technique, 36
      cross-over, 52
    post thrombotic state, 126
Iliocaval percutaneous phlebography, 39–41
Iliofemoral vein thrombosis, 126
Intraosseous phlebography, 42–48
  complications, 61–63
Iopamidol, osmolality, 55

Kidney tumours, caval involvement, 191
Klippel-Trenaunay syndrome, 58, 168–176
'Knot hole' effect, 72

Lea Thomas cannula, 42
Leg, lower, venous system, 8
  swollen, phlebographic investigation, 3
  veins, anatomical anomalies, 162–164
    deep, 9
    ectatic, 164
    peripheral, post thrombotic state, 123
    recanalisation, 124, 126, 128
Lumbar vein, ascending, 21
Lymph nodes, enlarged, iliocavogram in, 194

Malformations, venous, 152–179
  phlebography in, 3
Meglumine iothalamate contrast medium, 31, 55
  pain due to, 56
Metrizamide, osmolality, 55
  pain due to, 56

Needle for ascending phlebography, 27, 28

Os calcis injection, pain on walking after, 61
Osteomyelitis complicating intraosseous phlebography, 61
Ovarian veins, varicose, 214
Overcouch tube, 38
Overlapping shadows, 81

Pain complicating phlebography, 56
Palma's operation for iliac vein obstruction, 134, 135
Patient, position for ascending phlebography, 25
Pelvic
  varices, female, 213
  veins, deformities, causes, 187
    obstruction, 18, 19
Percutaneous iliocaval phlebography, 39
Perivenous pressure rise after deep vein thrombosis, 97
Peroneal veins, 8, 9, 13
  artefacts, 66, 67, 73, 82
  thrombosis, 100, 201, 202
  unfilled segment, 35
Phlebography
  ascending, techniques, 25–52
    modifications, 36
    standard, 33
    variations, 38
  complications, 54–65
    cardiac, 63
    local
  descending, 48–51
  during anticoagulant therapy, 201
  during thrombolytic therapy, 201
  dynamic, 3
  exercise, 38
  historical aspects, 1–5
  iliac cross-over, 52
  iliocaval, complications, 63
    percutaneous, 39
  in assessment of venous trauma, 208
  in clinical trials, 203
  in deep vein thrombosis, 85
  in extrinsic venous compression, 180, 182, 184
  in functional studies, 3
  in post thrombotic states, 119
  in primary varicose veins, 137
  in pulmonary embolism, 103
  in secondary varicose veins, 142
  in varicose veins, accuracy in demonstrating incompetent communicating veins, 148
  in venous malformations, 153
  indications, 2–4

INDEX 219

inferior vena cava, transcardiac, 41
intraosseous, 42–48
  complications, 61–63
  pubic, 213
  limitations, 4
  miscellaneous uses, 199–216
  perioperative, 199
  retrograde, 51
  superficial, 205
  tilt, 38
  tourniquets, 1, 26, 28, 32
  vena cava, inferior, transcardiac, 41
Phleboscopy, 1
Phlegmasia cerulea dolens, 100, 113, 114
Physiology of venous system, 6, 21–23
Plication, artefact due to, 82
Popliteal vein, 10–14
  artefacts, 68, 77
  thrombosis, 47, 100
Post thrombotic states, 116–136
  collateral pathways, 118, 119, 122, 123, 126, 128
  iliac veins, 126
  inferior vena cava, 126
  leg veins, peripheral, 123
  morbidity rates, 116
  phlebographic appearances, 122–132
  phlebography in, 119
Posture, effect on venous calibre, 77
Potts-Cournand needle, 39
Profunda femoris vein, 4, 12, 13, 30
Proteinuria in post thrombotic states, 130
Pubic phlebography, intraosseous, 213
Pudendal veins, 12, 13, 15
Pulmonary embolism
  complicating phlebography, 64
  fatal, site of origin, 106
  incidence, 101
  mortality rate, 103, 105
  phlebography in, 103
  recurrence, prevention, 104

Radiographic equipment, 31
Reactions to contrast media, 54
Recanalisation after thrombosis, 96
Rubber ankle tourniquet, 145

'Saphena-varix', 140
Saphenous vein
  long, 9, 13–15
  short, 9, 16
  incompetence, 140
  thrombosis, 203
  varicosities, 138–143, 166, 167
  Shadows, overlapping, 81
Sodium iothalamate contrast medium, 31
Streptokinase, 202
Supra-malleolar inflated cuff, 1

Techniques, ascending phlebography, 25–53
  modifications, 36
Television monitoring, 31
Thigh, communicating veins, 17

Thoracic veins, 45
Thrombectomy, phlebography during, 199
Thromboembolism, 85–115
Thrombogenesis, mechanism, 85
Thrombolytic therapy, phlebography during, 201
Thrombophlebitis, superficial, appearance, 88–90
Thrombosis
  complicating phlebography, 56
  deep vein, 3
    appearances, 90–100
    incidence, 101
    management, 112, 113
    phlebography in, 85, 106, 203
  iliofemoral, 210
  post thrombotic states, 116–136
  venous, 67
    appearances, 88–100
    differential diagnosis from venous artefact, 82
    phlebographic assessment, 207
Thrombus
  adherent, management, 105
  age, estimation, 101, 104
  'floating tail', 106, 107
  genesis, 85–88
  loose or adherent, 104
  management, 105
  'square cut', 106–107
Tibial veins, 8–10, 13
  compression, 197
Tilt phlebography, 38
Tourniquet, 1, 26, 28, 32
  causing pressure artefact, 81, 83
  pain due to, 56
Tunica adventitia, 6
  intima, 6
  media, 6

Ulcer, venous, 117, 119
  investigation, 2
Ureter, retrocaval or retroiliac, 158, 161
Urokinase, 202
Uterovaginal plexus, varicosities, 213

Valsalva manoeuvre, in ascending phlebography, 31
  risk of pulmonary embolism in, 64
Valves, aplasia, 167
  defects producing filling defects, 74, 75
  leg veins, in post thrombotic states, 123
  structure and function, 7
Varicography, 138, 204
Varicose veins, 137–152
  aetiology, 137
  in post thrombotic syndrome, 116
  ovarian, 214
  primary, 137–141
  recurrent, investigation of cause, 2
  secondary, 141–152
  uterovaginal plexus, 213
  vulval, phlebography, 210

220 INDEX

Veins
  abnormalities, artefactual, 67
  anatomical variations, 157–164
  aneurysms, 165, 166
  anterior tibial, 9, 10
  blood pressure in, 22
  blood supply, 21
  calf, 13, 17, 33
    artefacts, 68–70
    post thrombotic state, 96, 116
    thrombosis, phlebography in, 33, 209
  communicating, 17, 18
    incompetent, appearances, 147, 149, 150
    phlebography in, 2, 148
  compression
    extrinsic, 180–198
      anatomical causes, 186
      pathological causes, 187
      phlebography in, 180, 182, 184
    postural, 77
  digital, medial, as site for venepuncture, 26
  dysplasias, 153, 157
  ectatic, 164, 165
  epigastric inferior, 13, 20
  femoral, 23
    common, 13, 18, 35
      artefacts, 73, 77, 78
    deep, 12
    ligation, accidental, 200
    superficial, 7, 10–13, 15, 18, 23
      artefacts, 71, 72
      occlusion, post thrombotic, 122, 123
      pressures, post thrombotic, 134
      thrombosis, 204
  gastrocnemius, 12–14
  iliac
    common, 7, 15, 20, 44
      artefacts, 77–79
      compression defect, 125, 126, 180, 182–185
      filling, 35
      stenosis, 214
    deep circumflex, 20
    external, 12, 18–20
      aneurysm, 208
      artefacts, 70
    internal, 20
      obstruction, collaterals, 126, 128
        reconstructive surgery in, 134
      perforation during iliocaval phlebography, 63
      phlebography, bolus technique, 36
        cross-over, 52
      post thrombotic state, 126
  leg, anatomical anomalies, 162
    deep, 9
    ectatic, 164
    peripheral, post thrombotic state, 123
  lumbar, ascending, 21
  malformations, 153–179
    phlebographic investigation, 3
  obstruction, extrinsic, 3
    phlebography in, 2
  ovarian, varicose, 214
  pelvic, deformities, 187
    obstruction, 18, 19
  peroneal, 8, 9, 13
    artefacts, 66, 67, 73, 82
    thrombosis, 100, 201, 202
    unfilled segment, 35
  popliteal, 10–14
    artefacts, 68, 77
    thrombosis, 47, 100
  plantar, 9
  posterior arch, 15
  profunda femoris, 4, 12, 13, 30
  pudendal, 12, 13, 15
  renal, compression, 181
  saphenous
    long, 9, 13–15
    short, 9, 16
      incompetence, 140
      thrombosis, 203
    varicosities, 138–143, 166, 167
  soleal, 8, 13, 17
  structure, 6
  superficial, of lower leg, 10–15, 34
  thoracic, 45
  thrombosis, see Thrombosis.
  tibial, 8–10, 13
    compression, 197
  trauma, phlebographic assessment, 208
  ulceration, investigation, 2
  valves, see Valves.
  varicose, see Varicose veins.
Vena cava
  deformities, pathological, 186–189, 191–196
  inferior
    anatomical anomalies, 157–160
    compression, 180, 181
      in pregnancy, 187
    lower, 21
    obstruction, collateral pathways in, 130–132
    perforation during iliocaval phlebography, 63
    phlebography, bolus technique, 36
      transcardiac, 41
    post thrombotic state, 126
Venae comitantes, 10, 11, 94, 98
  as collateral pathways, 118, 122
Venepuncture for ascending phlebography, 27
Venous claudication, 126
  pressure, measurement, 132
    peripheral, measurement, 133
  system, anatomy, 6–21
    deep, demonstration of normality, 2
    investigation, non-invasive, 4
    physiology, 6, 21–23
Venturi effect, 76
Vulva, varices, drainage, 48
  phlebography, 210–212